Methods in Modern Biophysics

Springer
Berlin
Heidelberg
New York
Hong Kong
London
Milan
Paris
Tokyo

Bengt Nölting

Methods
in Modern Biophysics

 Springer

Dr. Bengt Nölting

Preussisches Privatinstitut für Technologie zu Berlin
Am Schlosspark 30
13187 Berlin

ISBN 3-540-01297-4 Springer Verlag Berlin Heidelberg New York

Cataloging-in-Publication Data applied for

Bibliographic information published by Die Deutsche Bibliothek
Die Deutsche Bibliothek lists this publication in the Deutsche Nationalbibliografie;
detailed bibliographic data is available in the Internet at <http://dnb.ddb.de>.

Springer-Verlag is a company in the specialist publishing group BertelsmannSpringer
@ Springer-Verlag Berlin Heidelberg 2004
Printed in Germany
http://www.springer.de

Coverdesign: Künkel & Lopka, Heidelberg

2/3020 UW   Printed on acid-free paper - 5 4 3 2 1 0 -

To   *Alan R. Fersht*
            Robert Huber
            Manfred Eigen
            Kurt Wüthrich

Preface

In the recent years we have seen a remarkable increase of the interest in biophysical methods for the investigation of structure-function relationships in proteins, cell organelles, cells, and whole body parts. Biophysics is expected to answer some of the most urgent questions: what are the factors that limit human physical and mental abilities, and how can we expand our abilities. Now a variety of new, faster and structurally higher-resolving methods enable the examination of the mysteries of life at a molecular level. Examples are X-ray crystallographic analysis, scanning probe microscopy, and nanotechnology. Astonishingly large molecular complexes are structurally resolvable with X-ray crystallography. Scanning probe microscopy and nanotechnology allow to probe the mechanical properties of individual biomolecules. Near-field optical microscopy penetrates Abbe's limit of diffraction and enables sub-200 nm resolution. Electron microscopy closes the gap between methods with molecular resolution and cellular resolution. Other methods, such as proteomics, mass spectrometry and ion mobility spectrometry, help us to study highly heterogeneous analytes and to understand extremely complex biological phenomena, such as the function of the human brain. Detailed mechanistic knowledge resulting from the application of these physical and biophysical methods combined with numerous interdisciplinary techniques will further aid the understanding of biological processes and diseases states and will help us to find rational ways for re-designing biological processes without negative side effects. This knowledge will eventually help to close the gap between humans and machines under consideration of all drawbacks, and to find cures for diseases and non-native declines of performance.

This book was mainly written for advanced undergraduate and graduate students, postdocs, researchers, lecturers and professors in biophysics and biochemistry, but also for students and experts in the fields of structural and molecular biology, medical physics, biotechnology, environmental science, and biophysical chemistry. The book is largely based on the lecture "Biophysical Methods" given by the author at the occasion of a visiting professorship at Vienna University of Technology. It presents a selection of methods in biophysics which have tremendously progressed in the last few years.

Chap. 1 introduces fundamentals of protein structures. Proteins have evolved to become highly specific and optimized molecules, and yet the class of proteins may be seen as the biomolecule class with the largest variety of functions. Surely the understanding of biological systems much depends on the understanding of protein structure, structure formation, and function. The next chapter (Chap. 2) presents important chromatographic methods for the preparation of proteins and other biomolecules. Many biophysical studies require this form of sample preparation and often a lot of time can be saved by using optimized procedures of sample purification. Mass spectrometry (Chap. 3) is important for the quality control in preparations of biomolecules, but also has a variety of further analytical applications. Chaps. 4–7 focus on methods for the chemical and structural characterization of biomolecules. X-ray crystallography (Sect. 4.1.2) probably offers the highest resolving power for large biomolecules and biological complexes, but it requires the preparation of high-quality crystals. Cheaper is infrared spectroscopy (Chap. 5) which may also comparably easily be applied in the fast time scale. Electron microscopy (Chap. 6) is particularly suitable for the structural resolution of complex biological systems at the size level of cells, cell organelles, and large molecular complexes. Different types of scanning probe microscopes (Chap. 7) can generate images of geometrical, mechanical, electrical, optical, or thermal properties of biological specimens with up to sub-nm resolution. In Chap. 8 (biophysical nanotechnology) we find novel methods for the mechanical characterization of individual biomolecules and for the engineering of novel nanotechnological structures and devices. The next two chapters (proteomics, Chap 9; and ion mobility spectrometry, Chap. 10) concentrate on two types of analytical methods for the characterization of complex samples such as human cells or bacteria. Finally Chap. 11 deals with some novel developments regarding the interaction of electromagnetic radiation with humans. Kinetics methods in biophysics were not much emphasized throughout the book since many of them can be found in the monograph "Protein Folding Kinetics" (Nölting, 1999b). The reader may refer to this monograph for more information on protein structure, transitions state theory in protein science, and on a variety of kinetic methods for the resolution of structural changes of proteins and other biomolecules.

Prof. Dr. Alan R. Fersht supported the development of a variety of modern biophysical methods in our extremely fruitful collaboration at Cambridge University. Prof. Dr. Robert Huber and Prof. Dr. Max F. Perutz initiated highly inspiring discussions regarding modern applications of protein X-ray crystallography. I am particularly indebted to Prof. Dr. Calvin F. Quate, Prof. Dr. Steven G. Sligar, and Prof. Dr. Joseph W. Lyding for an introduction into the AFM technology.

I am indebted to Prof. Dr. Joachim Voigt, Prof. Dr. Martin H. W. Gruebele, Prof. Dr. Kevin W. Plaxco, Dr. Gisbert Berger, and Dr. Min Jiang for proofreading the manuscript, and to Dr. Marion Hertel for processing the manuscript within Springer-Verlag.

Berlin, July 2003   Bengt Nölting

Contents

1 The three-dimensional structure of proteins 1
   1.1 Structure of the native state 1
   1.2 Protein folding transition states 9
   1.3 Structural determinants of the folding rate constants 12
   1.4 Support of structure determination by protein folding simulations 20

2 Liquid chromatography of biomolecules 23
   2.1 Ion exchange chromatography 23
   2.2 Gel filtration chromatography 28
   2.3 Affinity chromatography 31
   2.4 Counter-current chromatography and ultrafiltration 33

3 Mass spectrometry 37
   3.1 Principles of operation and types of spectrometers 37
      3.1.1 Sector mass spectrometer 38
      3.1.2 Quadrupole mass spectrometer 39
      3.1.3 Ion trap mass spectrometer 39
      3.1.4 Time-of-flight mass spectrometer 40
      3.1.5 Fourier transform mass spectrometer 43
      3.1.6 Ionization, ion transport and ion detection 44
      3.1.7 Ion fragmentation 45
      3.1.8 Combination with chromatographic methods 46
   3.2 Biophysical applications 49

4 X-ray structural analysis 59
   4.1 Fourier transform and X-ray crystallography 59
      4.1.1 Fourier transform 59
      4.1.2 Protein X-ray crystallography 69
         4.1.2.1 Overview 69
         4.1.2.2 Production of suitable crystals 69
         4.1.2.3 Acquisition of the diffraction pattern 71

      4.1.2.4 Determination of the phases: heavy atom replacement 76

      4.1.2.5 Calculation of the electron density and refinement 83

      4.1.2.6 Cryocrystallography and time-resolved crystallography 84

   4.2 X-ray scattering 85

      4.2.1 Small angle X-ray scattering (SAXS) 85

      4.2.2 X-ray backscattering 88

5 Protein infrared spectroscopy 91

   5.1 Spectrometers and devices 92

      5.1.1 Scanning infrared spectrometers 92

      5.1.2 Fourier transform infrared (FTIR) spectrometers 92

      5.1.3 LIDAR, optical coherence tomography, attenuated total 96
            reflection and IR microscopes

   5.2 Applications 102

6 Electron microscopy 107

   6.1 Transmission electron microscope (TEM)... 107

      6.1.1 General design 107

      6.1.2 Resolution 109

      6.1.3 Electron sources 110

      6.1.4 TEM grids 112

      6.1.5 Electron lenses 112

      6.1.6 Electron-sample interactions and electron spectroscopy 115

      6.1.7 Examples of biophysical applications 117

   6.2 Scanning transmission electron microscope (STEM) 118

7 Scanning probe microscopy 121

   7.1 Atomic force microscope (AFM) 121

   7.2 Scanning tunneling microscope (STM) 133

   7.3 Scanning nearfield optical microscope (SNOM) 135

      7.3.1 Overcoming the classical limits of optics 135

      7.3.2 Design of the subwavelength aperture 138

      7.3.3 Examples of SNOM applications 142

   7.4 Scanning ion conductance microscope, scanning thermal 143
        microscope and further scanning probe microscopes

8 Biophysical nanotechnology 147

   8.1 Force measurements in single protein molecules 147

8.2 Force measurements in a single polymerase-DNA complex 150
8.3 Molecular recognition 152
8.4 Protein nanoarrays and protein engineering 155
8.5 Study and manipulation of protein crystal growth 158
8.6 Nanopipettes, molecular diodes, self-assembled nanotransistors, ... 159
     nanoparticle-mediated transfection and further biophysical
     nanotechnologies

9 Proteomics: high throughput protein functional analysis 165
   9.1 Target discovery 166
   9.2 Interaction proteomics 168
   9.3 Chemical proteomics 172
   9.4 Lab-on-a-chip technology and mass-spectrometric array scanners 173
   9.5 Structural proteomics 174

10 Ion mobility spectrometry 175
   10.1 General design of spectrometers 175
   10.2 Resolution and sensitivity 180
   10.3 IMS-based "sniffers" 183
   10.4 Design details 184
   10.5 Detection of biological agents 193

11 Microwave auditory effects and the theoretical concept of thought 197
    transmission technology
   11.1 Theoretical concept 197
        11.1.1 Background of the concept of thought transmission technology .. 198
        11.1.2 Description of the technology 199
              11.1.2.1 Frequencies 200
              11.1.2.2 Radiation sources 201
   11.2 Examples of potential applications 202
        11.2.1 An automated device 202
        11.2.2 A small handheld device 203
        11.2.3 Biomedical microwave auditory and thought transmissions 204

12 Conclusions 211

References 213

Index 245

Symbols

→	arrow indicating a process or a coordinate axis	FTIR	Fourier transform infrared
⇢	arrow pointing to a label or indicating a distance	FTMS	Fourier transform mass spectrometer
Å	angström (10^{-10} m; 0.1 nm)	GC	gas chromatography
AC	alternating current	GPS	global positioning system
ADC	analog-to-digital converter	h	Planck constant (6.6261×10^{-34} J s)
AFM	atomic force microscope	HPLC	high pressure liquid chromatography
ATP	adenosine triphosphate		
BESSY	(Berlin Electron Synchrotron Storage Ring)	IHF	integration host factor
bp	base pair	IMS	ion mobility spectrometer
BSA	bovine serum albumin	IMU	inertial measurement unit
BSE	bovine spongiform encephalopathy	IR	infrared
°C	degree Celsius (kelvin − 273.15)	k_B	Boltzmann constant (1.3807×10^{-23} J K^{-1})
c	speed of light in vacuum (2.99792×10^8 m s^{-1})	KBr	potassium bromide
		kDa	kilodalton (kg mol^{-1})
CBMS	chemical-biological mass spectrometer	kJ	kilojoule (1 kJ = 240 cal)
CCD	charge coupled device	kp	kilopond (9.8066 N)
CD	circular dichroism	kV	kilovolt
CJD	Creutzfeldt-Jacob disease	l	liter (10^{-3} m^3)
CM	carboxy methyl	Laser	light amplification by stimulated emission of radiation
CNS	central nervous system	LD	linear dichroism
CsI	cesium iodide	LIDAR	light detection and ranging (measurement of light backscatter)
CTP	chain topology parameter		
Da	dalton (g mol^{-1})	μm	micrometer (10^{-6} m)
DC	direct current	MΩ	megaohm (10^6 V A^{-1})
dCTP	2'-deoxycytidine 5'-triphosphate	MALDI	matrix-assisted laser desorption ionization
DEAE	diethyl-amino-ethyl		
DNA	deoxyribonucleic acid	Maser	microwave amplification by stimulated emission of radiation
dsDNA	double-stranded DNA		
DTGS	deuterated triglycine sulfate	MCT	mercury cadmium telluride
e	elementary charge (1.6022×10^{-19} C)	m_e	electron rest mass (9.1094×10^{-31} kg)
eV	electron volt (1.6022×10^{-19} J)	ml	milliliter (10^{-6} m^3)
FPLC	fast performance liquid chromatography	mM	millimolar (6.0221×10^{20} liter^{-1})
		mol	6.0221×10^{23}

mV	millivolt (10^{-3} V)
M_w	molecular weight
m/z	mass-to-charge ratio
nA	nanoampere (10^{-9} A)
nm	nanometer (10^{-9} m)
NMR	nuclear magnetic resonance
NSOM	near-field scanning optical microscope – *see* SNOM
OCT	optical coherence tomography
ORF	open reading frame
pA	picoampere (10^{-12} A)
PCR	polymerase chain reaction
pg	picogram (10^{-12} g)
pI	isoelectric point
pN	piconewton (10^{-12} kg m s^{-2})
ppbv	part per billion volume (10^{-9})
PVC	polyvinyl chloride
PyMS	Py-MS, pyrolysis mass spectrometry
RADAR	radio detection and ranging
rms	root mean square
RNAse	ribonuclease
μs	microsecond (10^{-6} s)
SAXS	small angle X-ray scattering
SDOCT	spectral domain optical coherence tomography
SICM	scanning ion conductance microscope
SNOM	scanning near-field optical microscope
SPM	scanning probe microscope
ssDNA	single-stranded DNA
STEM	scanning transmission electron microscope
SThM	scanning thermal microscope
STM	scanning tunneling microscope
TEM	transmission electron microscope
TGS	triglycine sulfate
TIR	total internal reflection
TNT	trinitrotoluene
TOF	time-of-flight mass spectrometer
UV	ultra-violet
VIS	visible
VUV	vacuum ultra-violet

1 The three-dimensional structure of proteins

1.1 Structure of the native state

The human body contains the astonishing number of several 100,000 different proteins. Proteins are "smart" molecules each fulfilling largely specific functions such as highly efficient catalysis of biochemical reactions, muscle contraction, physical stabilization of the body, transport of materials in body fluids, and gene regulation. In order to optimally fulfill these functions, highly specific protein structures have evolved. The performance of humans, animals, and plants crucially depends on the integrity of these structures. Already small structural errors can cause diminishings of performance or even lethal diseases.

Proteins generally consist of thousands of atoms, such as hydrogen (H), carbon (C), nitrogen (N), oxygen (O), and sulfur (S). The van-der-Waals radii are about 1.4 Å for H, 2.1 Å for $-CH_3$, 1.8 Å for N, 1.7 Å for O, and 2.0 Å for S. Typical sizes of proteins range from a few nm to 200 nm. Since representations with atomic resolution of the whole protein molecule (Fig. 1.1a), or only its backbone (Fig. 1.1b), would be quite confusing for most proteins, it has become common to represent the protein structure as a ribbon of the backbone (Fig. 1.1c).

Multiple levels of structure are distinguished (see Nölting, 1999b): The most basic is the primary structure which is the order of amino acid residues. The 20 common amino acids found in proteins can be classified into 3 groups: non-polar, polar, and charged. Some physical properties of amino acids are given in Table 1.1. For the hydrophobicity of amino acids see Nölting, 1999b. A typical protein contains 50–1000 amino acid residues. An interesting exception is titin, a protein found in skeletal muscle, containing about 27,000 residues in a single chain. The next level, the secondary structure, refers to certain common repeating structures of the backbone of the polypeptide chain. There are three main types of secondary structure: helix, sheet, and turns. That which cannot be classified as one of these three types is usually called "random coil" or "other". Long connections between helices and strands of a sheet are often called "loops". The third level, the tertiary structure, provides the information of the 3-dimensional arrangement of elements of secondary structure in a single protein molecule or in a subunit of a protein molecule. The tertiary structure of a protein molecule, or of a subunit of a protein molecule, is the arrangement of all its atoms in space, without regard to its relationship with neighboring molecules or subunits. As this definition implies, a protein molecule can contain multiple subunits. Each subunit consists of only one polypeptide chain and possibly co-factors. Finally, the quaternary structure is the arrangement of subunits in space and the ensemble of its intersubunit contacts, without regard to the internal geometry of the subunits. The subunits in a quaternary structure are usually in noncovalent association. Rare exceptions are disulfide bridges and chemical linkers between subunits.

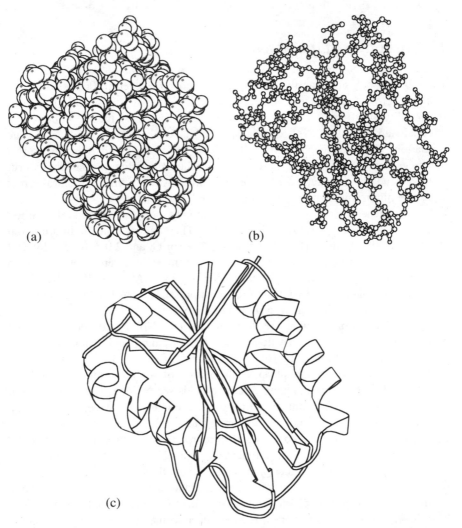

(a)   (b)

(c)

Fig. 1.1 The three-dimensional structure of the saddle-shaped electron transport protein flavodoxin from *Escherichia coli* (Hoover and Ludwig, 1997). (**a**) Space-fill representation of the complete molecule. (**b**) Ball-and-stick representation of the protein backbone. (**c**) Ribbon representation: ribbons, arrows, and lines symbolize helices, strands, and other, respectively. Coordinates are from the Brookhaven National Laboratory Protein Data Bank (Abola et al., 1997). The figure was generated using MOLSCRIPT (Kraulis, 1991)

Most proteins have only a marginal stability of $20-60\,\mathrm{kJ\,mol^{-1}}$ and can undergo conformational transitions (Nölting, 1999b). Small reversible conformational changes on a subnanometer scale occur very frequently. Reversible or irreversible molecular movements in the subnanometer or nanometer scale are essential for the function of many proteins. However, occasionally proteins irreversibly misfold into a non-native conformation. This can have dramatic consequences for the organism, especially when misfolded protein accumulates in the cell. A well known example of such a process is the misfolding of the prion protein (Figs. 1.2 and 1.3; Riek et al., 1996, 1998; Hornemann and Glockshuber, 1998). According to the "prion-only" hypothesis (Prusiner, 1999), a modified form of native prion protein can trigger infectious neurodegenerative diseases, such as Creutzfeldt-Jacob disease (CJD) in humans and bovine spongiform encephalopathy (BSE).

Table 1.1 Physical properties of natural amino acids

Amino acid	Molecular mass (Da)[a]	Partial molar volume $(\mathrm{cm^3\,mol^{-1}})$[b,c]	Partial molar volume of residue in protein $(\mathrm{cm^3\,mol^{-1}})$[b]
Alanine	89.09	60.4	54.7
Arginine	174.20	126.9	121.2
Asparagine	132.12	77.3	71.6
Aspartic acid	133.10	74.3	68.6
Cysteine	121.16	73.5	67.7
Glutamic acid	147.13	89.7	84.0
Glutamine	146.15	93.9	88.2
Glycine	75.07	43.2	37.5
Histidine	155.16	98.8	93.1
Isoleucine	131.17	105.6	99.9
Leucine	131.17	107.7	101.9
Lysine	146.19	111.4	105.7
Methionine	149.21	105.4	99.6
Phenylalanine	165.19	121.8	116.1
Proline	115.13	82.2	74.8
Serine	105.09	60.7	55.0
Threonine	119.12	76.9	71.1
Tryptophan	204.23	143.9	138.2
Tyrosine	181.19	123.7	118.0
Valine	117.15	90.8	85.1

[a] (Dawson et al., 1969; Richards, 1974; Coligan et al., 1996; Nölting 1999b)
[b] At 25 °C in water (Kharakoz, 1989, 1991, 1997)
[c] For the standard zwitterionic state

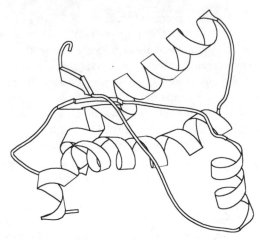

Fig. 1.2 Structure of the mouse prion protein fragment PrP(121–231) (Riek et al., 1996). The displayed secondary structure is strand$_1$ (128–131), helix$_1$ (144–153), strand$_2$ (161–164), helix$_2$ (172–194), helix$_3$ (200–224), coil (124–127, 132–143, 154–160, 165–171, 195–199). The figure was generated using MOLSCRIPT (Kraulis, 1991)

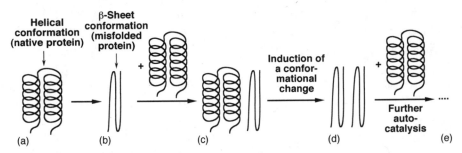

Fig. 1.3 A hypothetical mechanism of autocatalytic protein misfolding: with a low rate, the native helical conformation (**a**) spontaneously changes (misfolds) into a β-sheet conformation (**b**); contact of the misfolded protein with further correctly folded protein molecules (**c**) catalyzes further misfolding (**d, e**)

In soluble proteins, hydrophilic sidechains (that of aspartic acid, glutamic acid, lysine, arginine, asparagine, glutamine) have a higher preference for a location at the surface. Hydrophobic sidechains (that of alanine, valine, leucine, isoleucine, phenylalanine, tryptophan) are preferentially located inside the so-called hydrophobic core (Fig. 1.4). In contrast, the surface of membrane proteins often contains hydrophobic patches (Fig. 1.5).

Examples of the astonishing diversity of protein tertiary structure are shown in Figs. 1.6–1.8. Many proteins attain complicated multimeric structures. Fig. 6.18 in Chap. 6 shows an example of a complex assembly, the GroEL. For further details on the structures of proteins see Nölting, 1999b.

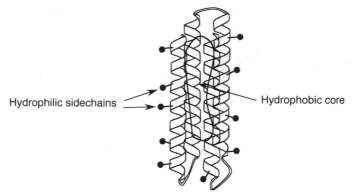

Fig. 1.4 In soluble proteins, charged and polar sidechains prefer a location at the surface. The sidechains of hydrophobic amino acids do not like to reside in an aqueous environment. That is why these sidechains are preferentially buried within the hydrophobic core

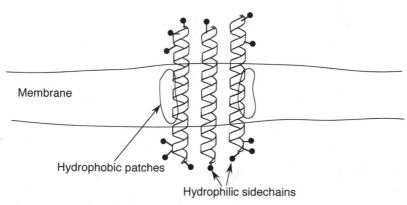

Fig. 1.5 Typical distribution of hydrophobic and hydrophilic sidechains in membrane proteins. The sidechains of hydrophobic amino acids are preferentially buried within the lipid portion of the membrane. Hydrophilic sidechains prefer contact with the bulk water outside the membrane

Next page: **Fig. 1.6** Examples of proteins with mainly helical secondary structure. **(a)** 1ACP: acyl carrier protein (Kim and Prestegard, 1990); **(b)** 1HBB: human hemoglobin A (Fermi et al., 1984); **(c)** 1BCF: iron storage and electron transport bacterioferritin (cytochrome b₁) (Frolow et al., 1994); **(d)** 1MGN: sperm whale myoglobin (Phillips et al., 1990); **(e)** 1QGT: assembly domain of human hepatitis B viral capsid protein (Wynne et al., 1999); **(f)** 2ABD: acyl-coenzyme A binding protein (Andersen and Poulsen, 1992); **(g)** 1FUM: the *Escherichia coli* fumarate reductase respiratory complex comprising the fumarate reductase flavoprotein subunit, the fumarate reductase iron-sulfur protein, the fumarate reductase 15-kDa hydrophobic protein, and the fumarate reductase 13-kDa hydrophobic protein (Iverson et al., 1999). Coordinates are from the Brookhaven National Laboratory Protein Data Bank (Abola et al., 1997). The figure was generated using MOLSCRIPT (Kraulis, 1991).

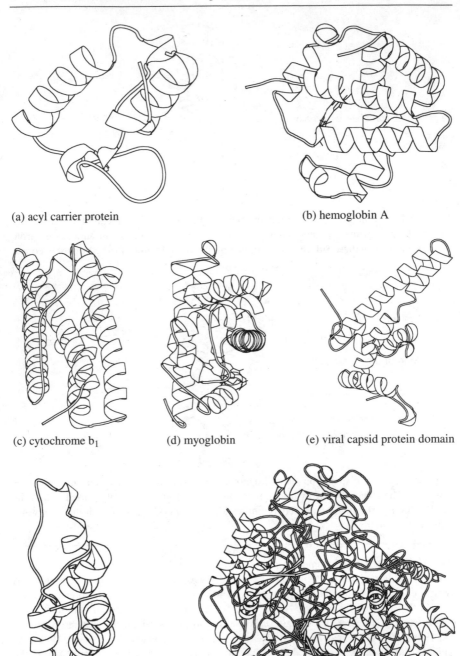

(a) acyl carrier protein

(b) hemoglobin A

(c) cytochrome b$_1$

(d) myoglobin

(e) viral capsid protein domain

(f) acyl-coenzyme A binding protein

(g) fumarate reductase respiratory complex

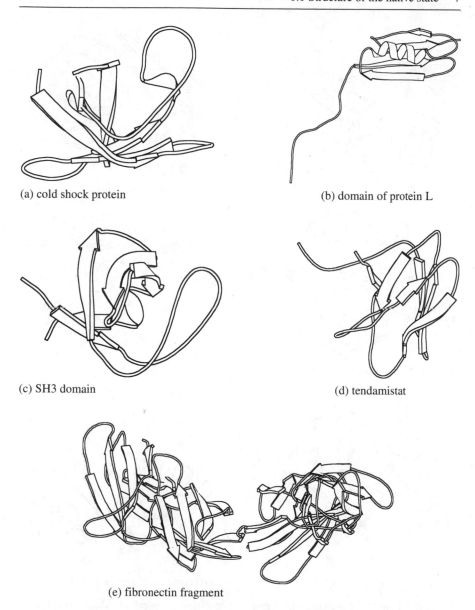

(a) cold shock protein

(b) domain of protein L

(c) SH3 domain

(d) tendamistat

(e) fibronectin fragment

Fig. 1.7 Examples of proteins with mainly sheet-shaped secondary structure. (**a**) 1CSP: major cold shock protein (CSPB) from *Bacillus subtilis* (Schindelin et al., 1993); (**b**) 2PTL: an immunoglobulin light chain-binding domain of protein L, (Wikström et al., 1995); (**c**) 1NYF: SH3 domain from fyn proto-oncogene tyrosine kinase (Morton et al., 1996); (**d**) 2AIT: α-amylase inhibitor tendamistat, (Kline et al., 1988); (**e**) 1FNF: fragment of human fibronectin encompassing type-III (Leahy et al., 1992). Coordinates are from the Brookhaven National Laboratory Protein Data Bank (Abola et al., 1997). The figure was generated using MOLSCRIPT (Kraulis, 1991)

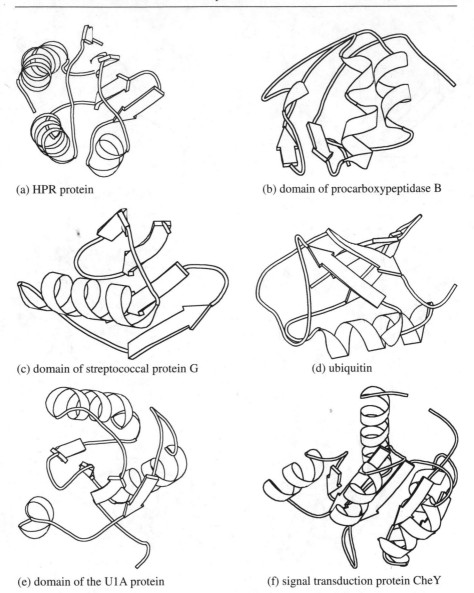

(a) HPR protein

(b) domain of procarboxypeptidase B

(c) domain of streptococcal protein G

(d) ubiquitin

(e) domain of the U1A protein

(f) signal transduction protein CheY

Fig. 1.8 Examples of proteins with significant amounts of helical and sheet-shaped structure. (**a**) 1HDN: histidine-containing phosphocarrier protein, (van Nuland et al., 1994); (**b**) 1PBA: activation domain from porcine procarboxypeptidase B, (Vendrell et al., 1991); (**c**) 1PGB: B1 immunoglobulin-binding domain of streptococcal protein G (Gallagher et al., 1994); (**d**) 1UBQ: human erythrocytes ubiquitin, (Vijay-Kumar et al., 1987); (**e**) 1URN: RNA-binding domain of the U1A spliceosomal protein complexed with an RNA hairpin, (Oubridge et al., 1994); (**f**) 3CHY: signal transduction protein CheY, (Volz and Matsumura, 1991). Coordinates are from the Brookhaven National Laboratory Protein Data Bank (Abola et al., 1997). The figure was generated using MOLSCRIPT (Kraulis, 1991)

1.2 Protein folding transition states

A considerable number of studies has been devoted to the resolution of folding
transition states, see, e.g., Nölting, 1999b. The structure of the folding transition

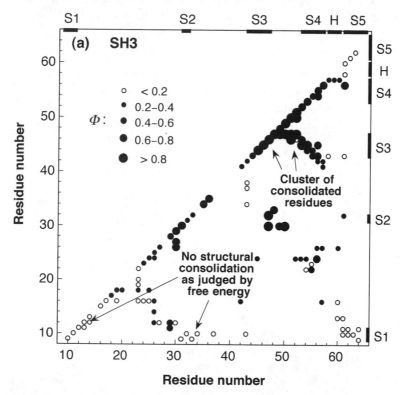

Fig. 1.9a Inter-residue contact map for the main folding transition state of the monomeric
protein src SH3 domain (Nölting and Andert, 2000). The sizes and fillings of the circles
indicate the magnitudes of structural consolidation, measured by the so-called Φ-value
(Nölting, 1999b). The diagonal of the plot displays secondary structure contacts, and
tertiary structure contacts are contained in the bulk of the diagram. Usually, high Φ-values
(large full circles) indicate a high degree of consolidation of structure and about native
interaction energies, and $\Phi \approx 0$ (small open circles) are diagnostic of little, if any,
formation of stable structure at the individual positions in the inter-residue contact space.
Moderate magnitudes of Φ ($\approx 0.2-0.8$) suggest different probabilities of the consolidation
of structure. Because of the possibility of the occurrence of non-native interactions in the
transition state, only clusters of several contacts (for Φ around 0.5 usually at least 5
contacts) may be used to draw statistically significant conclusions about the presence or
absence of a significant degree of structural consolidation. The positions of helices and
strands of β-sheets in the native state are indicated by bars, H1, H2, .. , and bars, S1, S2, .. ,
respectively

state is the structure of which formation represents the rate-limiting step in the folding reaction, i.e. the reaction of formation of the native conformation which usually starts with the unfolded polypeptide chain. Knowledge of transition state structures is important to understand the high efficiency of such folding reactions. The structures of many transition states of monomeric and also some dimeric and multimeric proteins provide evidence for a nucleation-condensation mechanism of folding in which structure growth starts with the formation of a diffuse folding nucleus which catalyzes further structure formation (Nölting, 1999b).

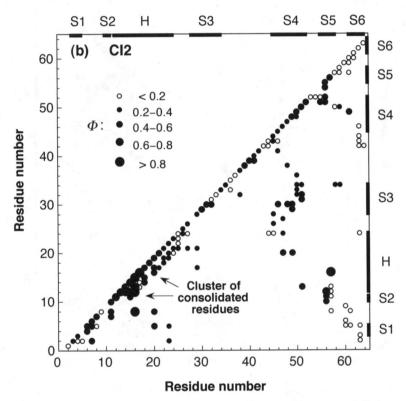

Fig. 1.9b Inter-residue contact map for the main folding transition state of chymotrypsin inhibitor 2 (CI2) (Nölting and Andert, 2000). The sizes and fillings of the circles indicate the magnitudes of structural consolidation, measured by the so-called Φ-value (Nölting, 1999b). *For further explanation see the legend for Fig. 1.9a on p. 9*

Fig. 1.9 a–d displays the structural consolidation of the transition states of four proteins. In these maps, the magnitudes of Φ-values are a measure or probability of structure formation at the corresponding locations in the inter-residue contact space. For example, large filled circles on the diagonal indicate consolidation of secondary structure contacts, and large filled circles in the bulk of the diagrams indicates consolidated tertiary structure contacts in the transition state (Nölting, 1998). The high structural resolution of the main transition states for the formation of native structure of these four small monomeric proteins (src SH3 domain, chymotrypsin inhibitor 2, barstar, barnase) and of the dimeric Arc repressor (not shown here) reveals that the most consolidated parts of each protein molecule in the transition state cluster together in the tertiary structure, and these clusters contain a significantly higher percentage of residues that belong to regular secondary structure than the rest of the molecule (Nölting and Andert, 2000). For many small monomeric and some dimeric proteins, the astonishing speed of protein folding can be understood as caused by the catalytic effect of the formation of clusters of residues which have particularly high preferences for the early formation of regular secondary structure in the presence of significant amounts of tertiary structure interactions (Nölting and Andert, 2000).

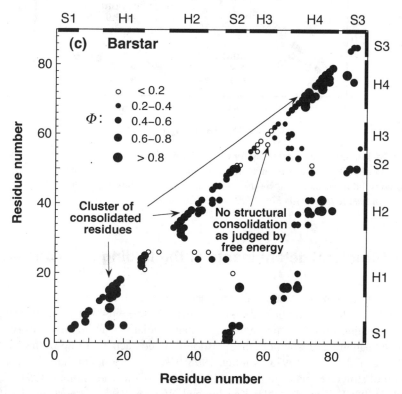

Fig. 1.9c Inter-residue contact map for the main folding transition state of barstar (Nölting and Andert, 2000). The sizes and fillings of the circles indicate the magnitudes of structural consolidation, measured by the so-called Φ-value (Nölting, 1999b). *For further explanation see the legend for Fig. 1.9a on p. 9*

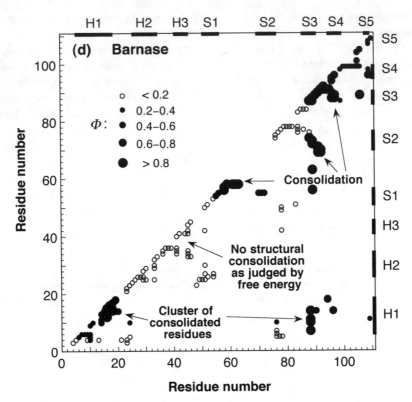

Fig. 1.9d Inter-residue contact map for the main folding transition state of barnase (Nölting and Andert, 2000). The sizes and fillings of the circles indicate the magnitudes of structural consolidation, measured by the so-called Φ-value (Nölting, 1999b). *For further explanation see the legend for Fig. 1.9a on p. 9*

1.3 Structural determinants of the folding rate constants

For the further understanding of the mechanism and extreme speed of protein folding, and for the rational design of artificial proteins and re-engineering of slowly-folding proteins with aggregating intermediates it is important to resolve, with subnanometer resolution, the question how contacts build up in the reaction (Nölting et al., 1995, 1997a; Nölting, 1998, 1999a, 1999b), and how this consolidation of structure relates to the speed of folding (Goto and Aimoto, 1991; Fersht et al., 1992; Dill et al., 1993; Karplus and Weaver, 1994; Orengo et al., 1994; Abkevich et al., 1995; Govindarajan and Goldstein, 1995; Hamada et al., 1995; Itzhaki et al., 1995; Nölting et al., 1995, 1997a; Fersht, 1995a, b; Gross, 1996; Kuwajima et al., 1996; Unger and Moult, 1996; Wolynes et al., 1996; Gruebele,

1999; Nölting, 1999b; Forge et al., 2000; Griko, 2000; Niggemann and Steipl, 2000; Nölting and Andert, 2000).

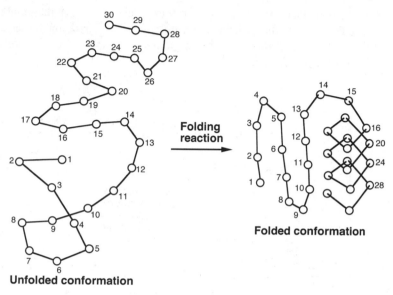

Fig. 1.10 Example for the formation of intramolecular contacts. Here the contacting residues in the folded conformation with the largest sequence separation are residues number 10 and 30. The set of distance separations in sequence between all contacting residues in space is called chain topology. It is an important determinant of the folding rate constant of the protein (Nölting et al., 2003)

One of the key questions is about the interplay between local and non-local interactions in the folding reaction (Tanaka and Scheraga, 1975, 1977; Gromiha and Selvaraj, 1997, 1999; Goto et al., 1999). In a number of studies it has been shown that the folding rate constants, k_f, of proteins depend on the contact order which is a measure of the complexity of the chain topology of the protein molecule (Fig. 1.10; Doyle et al., 1997; Chan, 1998; Jackson, 1998; Plaxco et al., 1998; Alm and Baker, 1999; Baker and DeGrado, 1999; Muñoz and Eaton, 1999; Riddle et al., 1999; Baker, 2000; Grantcharova et al., 2000; Koga and Takada, 2001). Proteins with a complicated chain topology, i.e. of which the native structure and the structure of the transition state contains many contacts of residues remote in sequence (Figs. 1.11 a, b; 1.12 a, b) have orders of magnitude lower folding rate constants, k_f, than proteins with a simple chain topology, i.e. of which the native structure and the structure of the transition state is dominated by contacts of residues near in sequence (Figs. 1.11 c, d; 1.12 c, d). Within the range of 10^{-1} s^{-1} $\leq k_f \leq 10^8$ s^{-1}, $-\log k_f$ correlates well with the so-called chain topology parameter, CTP, with a correlation coefficient of up to ≈ 0.87:

$$-\log k_{\mathrm{f}} \sim CTP\,, \qquad CTP = \frac{1}{L \cdot N} \sum \Delta S_{i,j}^2 \,, \qquad (1.1)$$

where L is the number of residues of the protein (chain length), N the number of inter-residue contacts in the protein molecule, $\Delta S_{i,j}$ the separation in sequence between the contacting residue number i and j, and "~" marks a linear correlation (Fig. 1.13; Nölting et al., 2003).

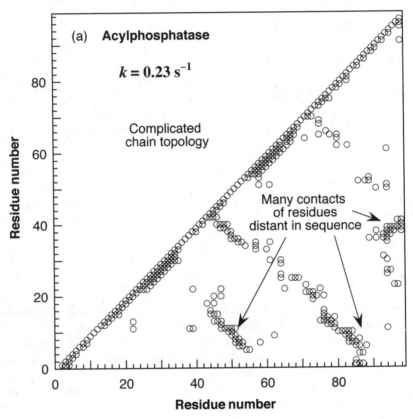

Fig. 1.11a Chain topologies (Nölting et al., 2003) of three proteins and a peptide with vastly different folding times: (**a**) acylphosphatase (Pastore et al., 1992), (**b**) FK506 binding protein (FKBP-12) (van Duyne et al., 1991), (**c**) λ-repressor dimer bound to DNA (Beamer and Pabo, 1992), and (**d**) the hairpin forming peptide from protein G (41–56) GEWTYDDATKTFTVTE (Achari et al., 1992; Muñoz and Eaton, 1999). Coordinates are from the Brookhaven National Laboratory Protein Data Bank (Abola et al., 1997). *Continued on the following pages*

The only important difference of the definition of *CTP* to the definition of the contact order is the quadratic dependence on $\Delta S_{i,j}$, and yet the fit is more stable and valid over a much larger range of rate constants and valid for both α-helix proteins and β-sheet proteins. The relation $-\log k_f \sim CTP$ can also reasonably well predict folding times of peptides. For various cut-off distances from 3.5 Å to 8.5 Å, the correlation coefficient, R, for $-\log k_f \sim CTP$ is 0.80–0.87 (Nölting et al., 2003; Fig. 1.14). Ignoring the inter-residue contacts involving hydrogen atoms which generally have less precisely known or fluctuating positions in the protein molecule causes only little if any effect on R (Nölting et al., 2003). When ignoring the data points for the small peptides, the R for $-\log k_f \sim CTP$ is still 0.75–0.81 for this range of cut-off distances.

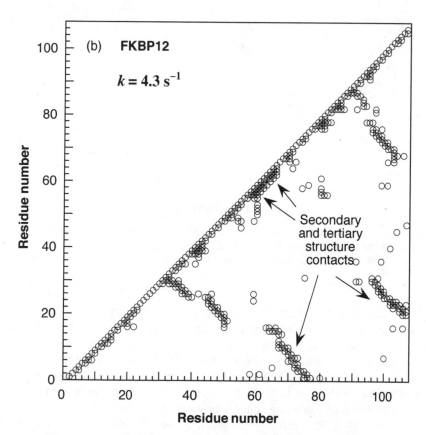

Fig. 1.11b Chain topology (Nölting et al., 2003) of FK506 binding protein (FKBP-12). Coordinates are from the Brookhaven National Laboratory Protein Data Bank (Abola et al., 1997). *For further chain topologies see pp. 14, 16, and 17*

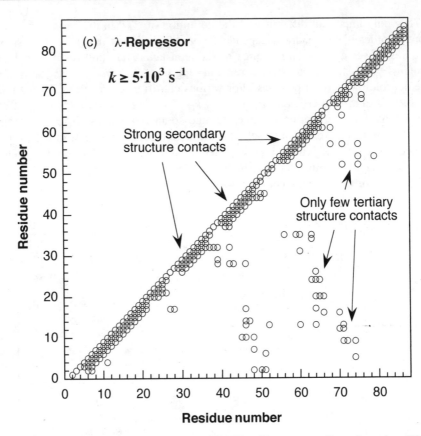

Fig. 1.11c Chain topology (Nölting et al., 2003) of λ-repressor dimer bound to DNA. Coordinates are from the Brookhaven National Laboratory Protein Data Bank (Abola et al., 1997). *For further chain topologies see pp. 14, 15, and 17*

A further important determinant of the speed of folding is the occurrence of some single strong interactions in the protein molecule. For example, some fast-folding proteins of thermophilic organisms contain a relatively large content of asparagine residues and salt bridges. These interactions can affect the rate of folding by a couple of orders of magnitude. $-\text{Log } k_f$ correlates also with the number of residues belonging to β-sheets. This may be due to the larger number of long-range secondary structure contacts in sheets than in helices.

The $-\log k_f \sim CTP$ is inconsistent with a zipper-like model for folding where the time of folding would be roughly proportional to the zipper length (sequence separation between zipper beginning and end). Obviously this relation is also inconsistent with a random-search mechanism where $-\log k_f \, [\text{s}^{-1}] \approx L - 9$.

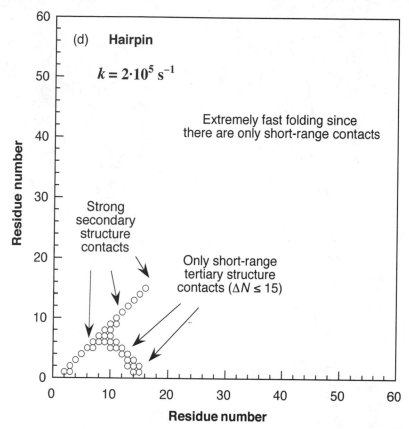

Fig. 1.11d Chain topology (Nölting et al., 2003) of the hairpin forming peptide from protein G (41–56) GEWTYDDATKTFTVTE. Coordinates are from the Brookhaven National Laboratory Protein Data Bank (Abola et al., 1997). *For further chain topologies see pp. 14–16*

The protein folding problem, i.e. the understanding of the astonishing speed, complexity and efficiency of folding (Nölting et al., 1995, 1997a; Nölting, 1999b; Nölting and Andert, 2000) has gained a large and still increasing importance in the context of folding-related diseases (Bellotti et al., 1998; Ironside, 1998; Brown et al., 1999; Gursky, 1999; Kienzl et al., 1999; Brown et al., 2000; Gursky and Alehkov, 2000), but also in the context of a variety of other exciting questions, such as macromolecular crowding inside the cell (Ellis and Hartl, 1999; van den Berg et al., 2000), high level expression of proteins (Hardesty et al., 1999; Kohno et al., 1999; Kramer et al., 1999), thermostability (Backmann et al., 1998; Williams et al., 1999) and packing problems (Efimov, 1998; Grigoriev et al., 1998, 1999; Efimov, 1999; Clementi et al., 2000a, 2000b).

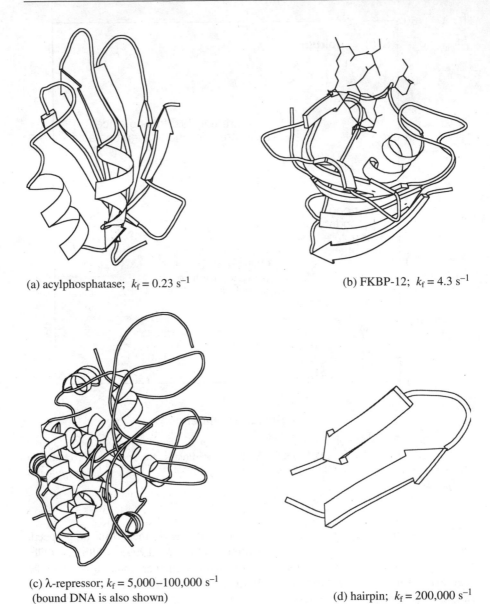

(a) acylphosphatase; $k_f = 0.23$ s^{-1}

(b) FKBP-12; $k_f = 4.3$ s^{-1}

(c) λ-repressor; $k_f = 5,000-100,000$ s^{-1}
(bound DNA is also shown)

(d) hairpin; $k_f = 200,000$ s^{-1}

Fig. 1.12 Structures of the three proteins and a peptide with vastly different folding rate constants, k_f: (**a**) acylphosphatase (Pastore et al., 1992), (**b**) FK506 binding protein (FKBP-12) (van Duyne et al., 1991), (**c**) λ-repressor dimer bound to DNA (Beamer and Pabo, 1992), and (**d**) the hairpin forming peptide from protein G (41–56) GEWTYDDATKTFTVTE (Achari et al., 1992; Muñoz and Eaton, 1999). Coordinates are from the Brookhaven National Laboratory Protein Data Bank (Abola et al., 1997). The figure was generated using MOLSCRIPT (Kraulis, 1991)

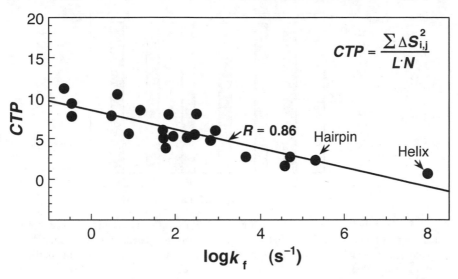

Fig. 1.13 The measured folding rate constants, k_f, of 20 proteins, a 16-residue β-hairpin and a 10-residue helical polyalanine peptide as a function of the chain topology expressed by the chain topology parameter, $CTP = L^{-1} N^{-1} \sum \Delta S_{i,j}^2$, where L is the number of residues of the macromolecule, N the total number of inter-residue contacts in the macromolecule, and $\Delta S_{i,j}$ the sequence separation between the contacting residues i and j (Nölting et al., 2003). The fit provides $\log k_f = 7.56 - 0.895 \cdot CTP$ with a correlation coefficient of 0.86. Within the range of 10^{-1} s^{-1} ≤ k_f ≤ 10^8 s^{-1}, predictions of the folding rate constants of peptides and proteins are accurate to typically a couple of orders of magnitude. The relation between structure and rate of folding is so important because it tells us a lot about the mechanism of protein folding and helps to solve the so-called folding paradox (see Nölting, 1999b, Nölting et al., 2003). Inter-residue contacts were calculated at a cut-off distance of 4 Å, and no contacts of hydrogen atoms were included in the calculations. Coordinates of the proteins and the β-hairpin were taken from the Brookhaven National Laboratory Protein Data Bank (Abola et al., 1997). For the choice of coordinates see Nölting et al., 2003. Coordinates of the 10-residue helical polyalanine peptide were calculated with the program FoldIt (Jésior et al., 1994). 18 rate constants from ref. (Jackson, 1998) and the k_f of the 16-residue β-hairpin were chosen as previously selected in ref. (Muñoz and Eaton, 1999). The k_f of the 10-residue helical polyalanine peptide was estimated using data in (Williams et al., 1996; Gruebele, 1999; Zhou and Karplus, 1999; Nölting, 1999b). Embedded in a lipid membrane, similar helices in folded proteins undergo intense vibrations with a frequency of 10^7 s^{-1} and several 0.1 Å elongation (e.g., Voigt and Schrötter, 1999). The k_f for the thermostable variant of λ-repressor and for the engrailed homeodomain, ≈50,000 s^{-1}, and 37,000 s^{-1} are from (Burton et al., 1996, 1997), and (Mayor et al., 2000), respectively (Nölting et al., 2003)

Studies on protein folding have contributed to the better understanding of hydrophobic interaction (Drablos, 1999; Garcia-Hernandez and Hernandez-Arana, 1999; Chan, 2000; Czaplewski et al., 2000), hydrophilic interaction (Jésior, 2000),

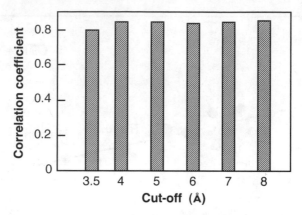

Fig. 1.14 Correlation coefficient for $-\log k_f \sim CTP$ for different cut-off distances for the calculation of the contacts, as indicated. No contacts of hydrogen atoms were included in the calculations. Including these contacts leads to a slightly higher correlation coefficient (Nölting et al., 2003)

charge interaction (Åqvist, 1999; de Cock et al., 1999), sidechain association (Galzitskaya et al., 2000), and disulfide formation (Chang et al., 2000a, 2000b).

Speeding up folding was achieved by design of sequences with good folding properties (Irbäck et al., 1999) and facilitating folding with helper molecules, so-called chaperones (Csermely, 1999; El Khattabi et al., 1999; Itoh et al., 1999; Kawata et al., 1999; Yamasaki et al., 1999; Gutsche et al., 2000a, 2000b), and taking carbohydrates as templates for *de novo* design of proteins (Brask and Jensen, 2000).

Protein folding has gained interest also regarding RNA folding energy landscapes (Chen and Dill, 2000), the interpretation of multi-state kinetics (Bai, 1999, 2000; Goldbeck et al., 1999), interpretation of DSC data towards cooperative formation of a folding nucleus (Honda et al., 1999; Honda et al., 2000), the evolution of structure formation (Chan, 1999; D'Alessio, 1999a, 1999b), protein secretion (Chambert and Petit-Glatron, 1999; Berks et al., 2000), and protein structure prediction (Crawford, 1999).

1.4 Support of structure determination by protein folding simulations

Theoretically, the structure of the native state of a protein can be determined by calculating the energies of all conformations of the molecule. This is true even if the native conformation does not correspond to the global energy minimum. For example, with a few additional experimentally obtained distance constraints one could decide which is the native structure. Unfortunately, the number of possible conformations of a polypeptide chain is astronomically large. For example, as judged by the entropy, for a protein comprising 100 residues it is of the order of 10^{100} (Nölting, 1999b). There are some more optimistic estimates which are based on mechanistic considerations, but still the number of conformations is astronomically large. A further problem is that there are large positive and negative contributions to the protein stability: The stability of the molecule is given by the difference of two large almost equal numbers (Nölting, 1999b). In order to calculate the global energy minimum or a folding pathway with sufficient precision, these two numbers would need to be known with about 3–4 significant digits. Currently the theory of molecular energies is not precise enough to meet this requirement. That is why it has not yet been possible to calculate the global energy minimum of an average-sized protein without significant approximations and profound simplifications. Only recently, groundbreaking molecular dynamics simulations on a 23-residue mini-protein found the energy minimum in 700 μs of simulation (Snow et al., 2002).

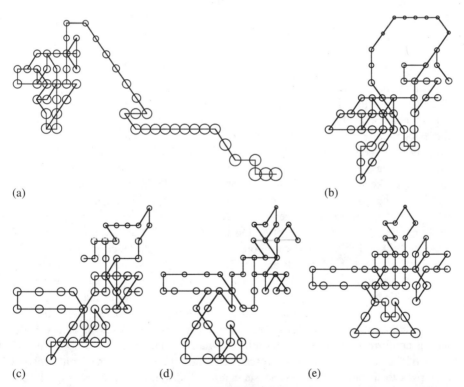

(a)   (b)

(c)   (d)   (e)

Fig. 1.15 Support of structure determination by simulation of protein folding. (**a**) Step 100 of the simulation: initial collapse to a non-native conformation. (**b**) Step 400: formation of a molten-globule-like state. (**c**) Step 4,480 and (**d**) step 17,990: further condensation and reorganization of the molten-globule intermediate. (**e**) Step 38,174: formation of a native-like state. Each circle represents an amino acid residue of the protein

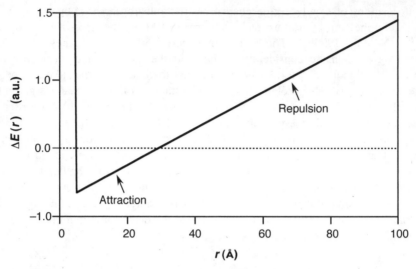

Fig. 1.16 Hydrophobic potential used for the folding simulation shown in Fig. 1.15

Due to their extreme simplicity, lattice models for the protein structure and statistical energies have become especially prominent (see, e.g., Shakhnovich et al., 1996; Shakhnovich 1997; Mirny and Shakhnovich, 2001). In these models, often the amino acid residues are represented by spheres and the possible angles of the backbone are significantly restricted, e.g., only 0° and ±90° are allowed. Surprisingly, these simple approaches often yield reasonable results.

Fig. 1.15 exemplary shows lattice simulations which could fold small proteins into native-like structures. The hydrophobic potential used for these simulations is similar to the potential described by Casari and Sippl (1992), but has a strong repulsion at very short distances (Fig. 1.16). For the attractive component, the same relative factors for pairs of amino acids were used as given by Casari and Sippl (1992) in Table 2. The start conformations are random combinations of the structural elements helix, sheet and random coil. The use of not purely random start conformations, but start conformations that contain fluctuating secondary structure elements speeds up the simulation by several orders of magnitude. The aim was not to calculate a unique native structure, but is to find a set of low-energy conformations. Experimental constraints are then used to rule out the wrong conformations and to determine the native conformation. Important features of the folding reaction are resembled: the initially expanded conformation collapses to a molten-globule-like state after 400 simulation steps (Fig. 1.15b) which reorganizes after a total of 38,174 simulation steps to a native-like conformation (Fig. 1.15e).

2 Liquid chromatography of biomolecules

Proteins, peptides, DNA, RNA, lipids, and organic cofactors have various characteristics such as electric charge, molecular weight, hydrophobicity, and surface relief. Purification is usually achieved by using methods that separate the biomolecules according to their differences in these physical characteristics, such as ion exchange (Sect. 2.1), gel filtration (Sect. 2.2), and affinity chromatography (Sect. 2.3).

2.1 Ion exchange chromatography

In ion exchange chromatography, the stationary solid phase commonly consists of a resin with covalently attached anions or cations. Solute ions of the opposite charge in the liquid, mobile phase are attracted to the ions by electrostatic forces. Adsorbed sample components are then eluted by application of a salt gradient which will gradually desorb the sample molecules in order of increasing electrostatic interaction with the ions of the column (Figs. 2.1–2.3). Because of its excellent resolving power, ion exchange chromatography is probably the most important type of chromatographic methods in many protein preparations.

The choice of ion exchange resin for the purification of a protein largely depends on the isoelectric point, pI, of the protein. At a pH value above the pI of a protein, it will have a negative net charge and adsorb to an anion exchanger. Below the pI, the protein will adsorb to a cation exchanger. For example, if the pI is 4 then in most cases it is advisable to choose a resin which binds to the protein at a pH > 4. Since at pH > 4 this protein is negatively charged, the resin has to be an anion ion exchanger, e.g., DEAE. One could also use a pH < 4 and a cation exchanger, but many proteins are not stable or aggregate under these conditions. If, in contrast, the protein we want to purify has a pI = 10, it is positively charged at usually suitable conditions for protein ion exchange chromatography, i.e. at a pH around 7. Thus, in general for this protein type we have to choose a cation ion exchange resin, e.g., CM, which is negatively charged at neutral pH.

The capacity of the resin strongly depends on the pH and the pI of the proteins to be separated (Fig. 2.4; Table 2.1), but also on the quality of the resin, the applied pressure, and the number of runs of the column (Fig. 2.5). To improve the life of the resin, it should be stored in a clean condition in the appropriate solvent and not be used outside the specified pH range and pressure limit.

For the separation of some enzymes which may lose their activity by contact with metals in the wall of stainless steel columns, glass-packed columns may be more appropriate. The chromatographic resolution mainly depends on the type of biomolecules, type and quality of the resin, ionic strength gradient during elution, temperature, and the geometry of the column.

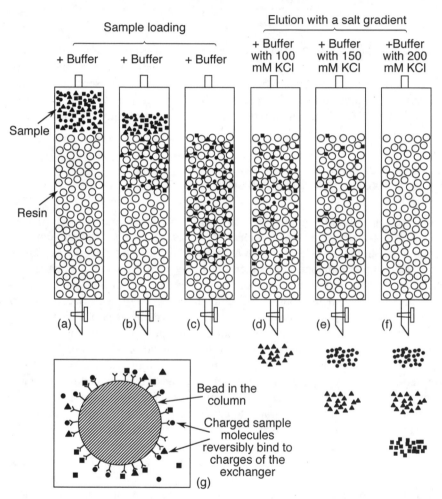

Fig. 2.1 Example of ion exchange chromatography. (**a**)–(**c**) Loading the column: mobile anions (or cations) are held near cations (or anions) that are covalently attached to the resin (stationary phase). (**d**)–(**f**) Elution of the column with a salt gradient: the salt ions weaken the electrostatic interactions between sample ions and ions of the resin; sample molecules with different electrostatic properties are eluted at different salt concentrations, typically between 0–2 M. (**g**) Interaction of sample molecules with ions attached to the resin: at a suitable pH and low salt concentration, most of the three types of biomolecules to be separated in this example reversibly bind to the ions of the stationary phase

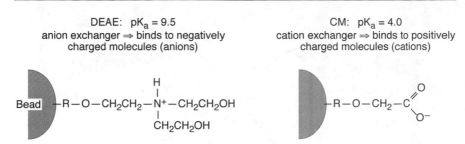

DEAE: $pK_a = 9.5$
anion exchanger $\Rightarrow$ binds to negatively
charged molecules (anions)

CM: $pK_a = 4.0$
cation exchanger $\Rightarrow$ binds to positively
charged molecules (cations)

Fig. 2.2 Two ion exchangers: diethyl-amino-ethyl (DEAE) and carboxy methyl (CM). The positive charge of DEAE attracts negatively charged biomolecules. CM is suitable for purification of positively charged biomolecules

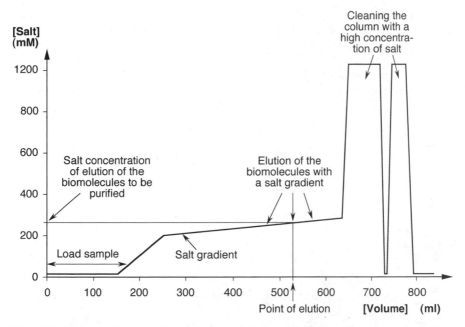

Fig. 2.3 Example for the salt concentration during adsorption of a sample to an ion exchange column, subsequent elution of the sample, and cleaning of the column. Example of a purification protocol: First the solution of biomolecules and impurities in buffer contained in a syringe is loaded onto the column. The biomolecules and some of the impurities bind to the ions attached to the resin. Loading is completed and non-binding molecules are partly rinsed through the column with some further buffer. The next step is to apply a salt gradient with a programmable pump which mixes buffer with extra salt-containing buffer. The steep salt gradient at the beginning elutes most of the weakly binding impurities. At a certain salt concentration, the biomolecules to be purified elute from the column. Elution is monitored with an absorption detector at 280 nm wavelength and the sample fraction collected. After each run the column is cleaned with 1–2 M KCl. This removes most of the strongly binding sample impurities

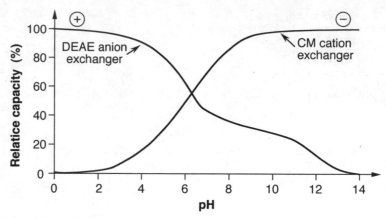

Fig. 2.4 Charge properties of anion and cation exchangers. DEAE has a significant capacity at low and medium pH; CM is highly capacious at high and medium pH

Table 2.1 Properties of some important ion exchangers

Functional group	Type of exchanger	pH range
$-\overset{\backslash}{\underset{/}{N^+}}-CH_3$	Quaternary amine (strong anion)	1 – 11
$-NH_2$	Primary amine (weak anion)	1 – 8
$-NH$	Secondary amine (weak anion)	1 – 7
$-\overset{\backslash}{\underset{/}{N}}$	Tertiary amine (weak anion)	1 – 6
$-COO^-$	Carboxylic acid (weak cation)	6 – 14
$-SO_3^-$	Sulfonic acid (strong cation)	1 – 14

The experimental set-up (Fig. 2.6) often just consists of a bottle with buffer, a bottle with buffer with salt, a programmable FPLC or HPLC pump, the column, a detector and recorder of absorption at 280 nm, or occasionally at 220 nm, and a sample collector. If the right conditions for protein preparation are unknown, a pre-run is performed with a small fraction of the sample. Attention should be paid not to overload the column in preparative runs since this can shift peak positions and lead to substantial sample losses. In many cases of modern high expression of recombinant proteins, it is possible to obtain a protein with 99% purity with a

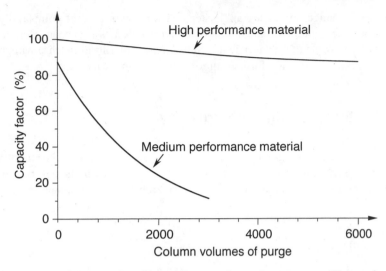

Fig. 2.5 Change of the capacity of ion exchange columns due to usage. High performance columns operated at the appropriate pressure and pH can last many 1000 runs

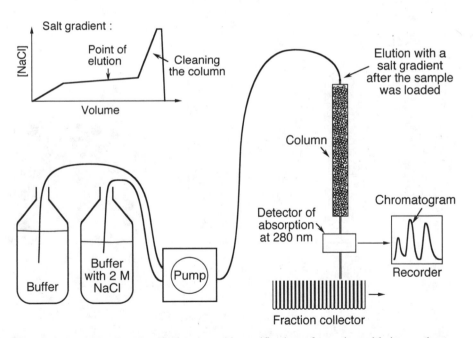

Fig. 2.6 Typical setup for chromatographic purification of proteins with ion exchange FPLC. The pump mixes the salt gradient for sample elution after the sample was loaded, e.g., with a syringe

single ion exchange chromatographic step. However, in case of comparably low expression levels and substantial sample contamination, ion exchange chromatography alone may not be sufficient. Subsequent gel filtration chromatography (Sect. 2.2) can significantly further improve the protein purity.

2.2 Gel filtration chromatography

This type of chromatography is a variant of size exclusion chromatography (molecular exclusion chromatography), and is also known as gel permeation chromatography. It lacks an attractive interaction between the stationary phase

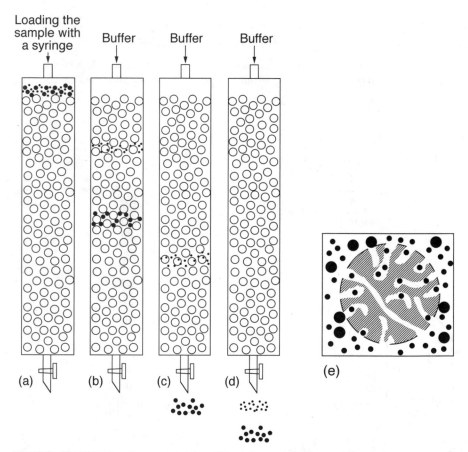

Fig. 2.7 Gel filtration chromatography. When the sample passes through the porous gel, small sample molecules can enter the pores, causing them to flow slower through the column. Large molecules which cannot enter the pores, pass through the column at a faster rate than the smaller ones. Correct pore sizes and solvents are crucial for a good separation

(gel) and solute. The sample solution passes through the porous gel separating the molecules according to their size. The smallest molecules enter the bead pores, resulting in a relatively long flow path and long retention. Large molecules cannot enter the pores and have to flow around them, resulting in a relatively short flow path (Figs. 2.7–2.10).

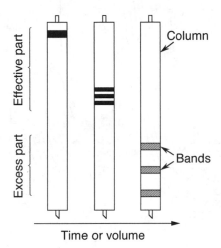

Fig. 2.8 Band broadening in a column with too long a geometry. The so-called effective part of the column is sufficient for separation. Excessively long columns do not improve purity, but just cause dilution of the sample by band broadening

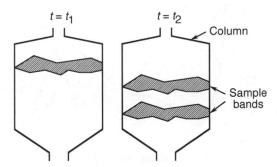

Fig. 2.9 Band broadening in a column with too large a diameter. Despite the column length is about right to separate the two bands, significant sample dilution and possibly contamination occurs due to inhomogeneous loading of the column

Gel filtration chromatography is also an auxiliary method for assessing the molecular weight of biomolecules (Fig. 2.11). Although there are more precise methods, e.g., mass spectrometry (see Chap. 3), gel filtration chromatography is important for the measurement of monomer-multimer equilibria at about μM-concentrations of biomolecules.

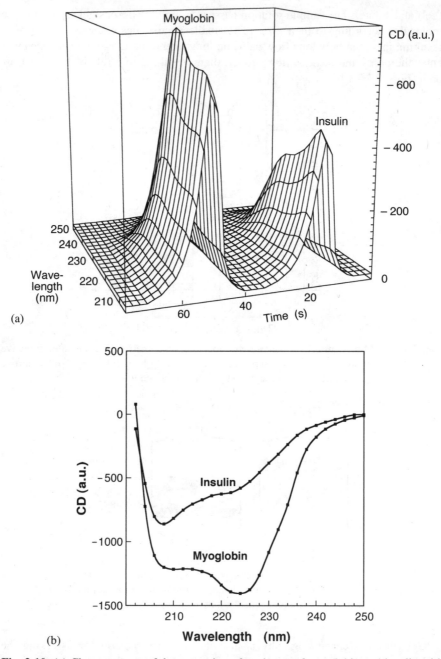

Fig. 2.10 (a) Chromatogram of the separation of a mixture of myoglobin and insulin with multichannel circular dichroism (CD) detection. The multiplex advantage of the multichannel detection prevents distortion of the shape of the spectra (see Nölting, 1999b). (b) CD spectra of myoglobin and insulin for comparison

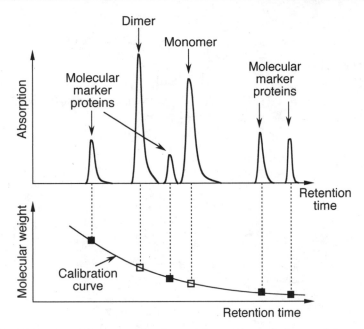

Fig. 2.11 Molecules of known molecular weight enable an estimate of the molecular weight of the unknown molecule. In this case, two peaks of the investigated molecule indicate a monomer-dimer equilibrium

2.3 Affinity chromatography

Affinity chromatography is a method enabling purification of biomolecules and other macromolecules with respect to individual structure or function. It utilizes the highly specific binding of the macromolecule to a second molecule which is attached to the stationary phase. The principle of operation is as follows: (a) the sample is injected into the column; (b) buffer is rinsed through the column, so that sample molecules with no affinity to the stationary phase are eluted from the column, but sample molecules with a high affinity for the stationary phase are retained in the column; (c) the retained sample molecules are eluted from the column by buffer with a high salt concentration or a different pH or a different solvent composition (Fig. 2.12). The preparation of the resin can be performed by using a number of protein tags. The tags should not cause artificial interactions and should not alter the conformation of the tagged protein. Very common are poly-histidine tags that are attached to the protein by genetic engineering (Fig. 2.13). The tag typically consists of 8–12 histidine residues. It binds to nickel compounds at the surface of the chromatography beads. Fig. 2.14 illustrates a somewhat different variant of affinity chromatography in which misfolded proteins are continuously refolded by chaperones and eluted with buffer.

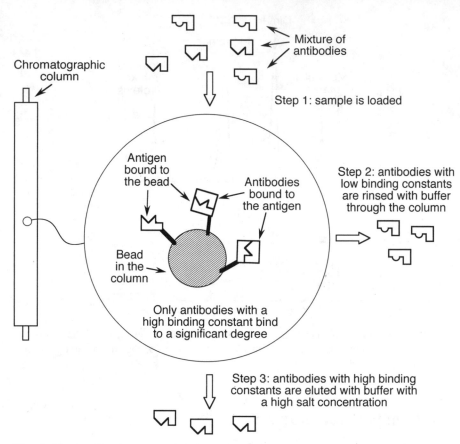

Fig. 2.12 Purification of antibodies with affinity chromatography: The antigen is chemically bound to the beads of the column and the mixture of antibodies is rinsed through the column. Antibodies with high binding constants bind to the antigen and are eluted later with a buffer with a high salt concentration

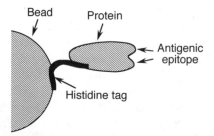

Fig. 2.13 Attachment of a protein to a bead of an affinity column with a histidine tag. About 10 histidine residues were attached to the protein by genetic engineering, e.g., by polymerase chain reaction (PCR) mutagenesis (see, e.g., Nölting, 1999b). The histidine residues strongly bind to the bead made from a nickel chelate resin

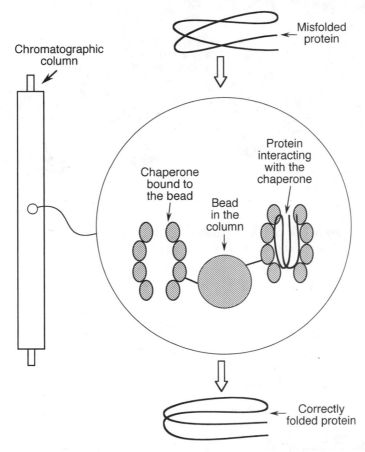

Fig. 2.14 Refolding of expensive, poorly folding proteins: Folding chaperones, also known as chaperonins, are attached to the beads and the unfolded or misfolded protein is rinsed through the column. The chaperone interacts with the sample protein and catalyses its folding into the correct conformation

2.4 Counter-current chromatography and ultrafiltration

A relatively old method of chromatography is the Craig counter-current distribution apparatus (Fig. 2.15). Nowadays it serves for the large-scale purification of some chemicals for which other chromatographic methods are too expensive. As in other types of counter-current chromatography, both stationary and mobile phase are liquids and separation is based on sample partition between the two liquids. It may, e.g., function as follows (Fig. 2.15): (a) A certain biochemical has a higher solubility in phase A than impurities of the biochemical, but has a lower solubility in phase B than the impurities. (b) Phase B with a high concentration

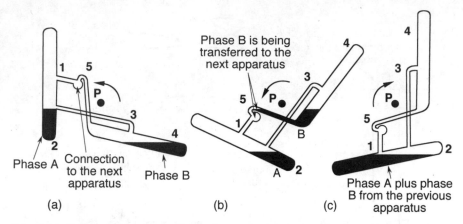

Fig. 2.15 Craig counter-current distribution apparatus: both stationary and mobile phases are liquids. Sample separation is based on its partition between the two liquid phases (see text)

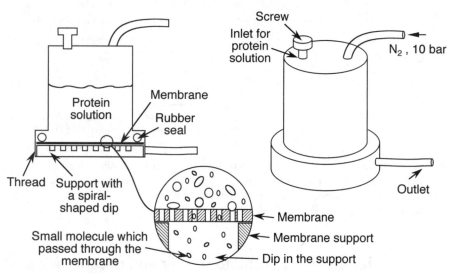

Fig. 2.16 Ultrafiltration device (supplied, e.g., by Amicon Inc., Beverly, MA). Pressurized nitrogen from a nitrogen flask presses the protein solution against the membrane. Small molecules pass the membrane and are collectable at the outlet. Large molecules stay in the ultrafiltration vessel

of impurities is transferred to the next apparatus and fresh phase B is transferred from the previous apparatus to the shown apparatus. (c) Phases A and B are mixed and separated again, and the process continues with step (a). During suc-

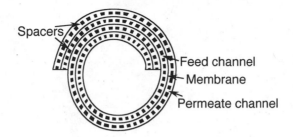

Spacers

Feed channel

Membrane

Permeate channel

Fig. 2.17   Side view of a spiral cartridge concentrator (e.g., Millipore Corporation, Bedford, MA).   Pressure is applied by centrifuging the concentrator. Similarly to the pervious ultrafiltration device (Fig. 2.16), small molecules pass the membrane and large molecules are retained

cessive cycles, different chemicals move though a chain of counter-current distribution apparatuses with different speeds, and are collected, e.g., at the end of the chain.

Strictly speaking, ultrafiltration (Figs. 2.16 and 2.17) is not a chromatographic method.   However, it should be mentioned here since it is an extremely useful tool of sample preparation prior to chromatography and can sometimes even substitute chromatography.   It is applicable for (a) protein purification, (b) buffer exchange, and (c) concentrating protein solutions.   Purification of a protein with a particular molecular weight, M_w, requires two steps: (a) First, one runs the ultrafiltration apparatus with a membrane with a cut-off higher than M_w and collects the solution leaving the vessel.   (b) Then, one runs the apparatus with a membrane with a cut-off lower than M_w and collects the solution remaining in the vessel.

3 Mass spectrometry

Mass spectrometry is an incredibly important analytical technique for the identification of molecules by way of measuring their mass-to-charge ratios, m/z, in the ionized state. It is particularly useful for the detection and analysis of traces of macromolecules down to less than 1 pg (10^{-12} g). The general design of a mass spectrometer comprises sample injector, sample ionizer, mass analyzer and ion detector (Fig. 3.1). First the sample is injected into the ionizer which ionizes sample molecules. Then sample ions are analyzed and detected. To prevent collisions with gas molecules, sample ionizer, mass analyzer and ion detector are generally operated in vacuum.

Fig. 3.1 General design of a mass spectrometer

The ion separation power of mass spectrometers is described by the resolution, R, defined as:

$$R = \frac{m}{\Delta m} \; , \tag{3.1}$$

where m and Δm are the ion mass and mass difference between two resolvable peaks in the mass spectrum, respectively. R typically ranges between 100 and 500,000.

3.1 Principles of operation and types of spectrometers

According to their mass analyzer designs, there are five important types of mass spectrometers (MS): (a) magnetic and/or electric sector MS (Figs. 3.2 and 3.3), (b) quadrupole MS (Fig. 3.4), (c) ion trap MS (Fig. 3.5), (d) time-of-flight MS (Figs. 3.6–3.9), and (e) Fourier transform MS (Fig. 3.10). Time-of-flight mass spectrometers (TOFs) often are less expensive than other types of mass spectrometers and have, compared to quadrupole MS and many sector MS, the advantage of recording the masses of all ions injected into the analyzer without scanning, contributing to a high sensitivity. TOFs usually have a smaller mass range and resolving power than Fourier transform mass spectrometers (FTMS).

3.1.1 Sector mass spectrometer

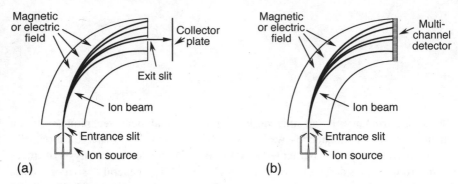

Fig. 3.2 Single magnetic or electric sector mass spectrometer with a single channel (**a**) and multichannel (**b**) detector, respectively. Ions leaving the ion source are accelerated and passed through the sector in which the electric or magnetic field is applied perpendicular to the direction of the ion movement. The field bends the ion flight path and causes ions with different m/z to travel on different paths. In scanning mass analyzers (**a**) the electric or magnetic field strength is varied and only one mass detected at a time. In non-scanning mass analyzers (**b**) all masses are recorded simultaneously within a limited mass range with the help of a multichannel detector

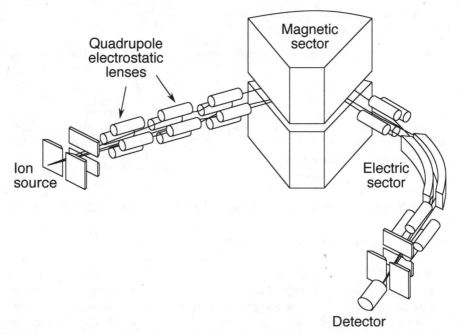

Fig. 3.3 Advanced virtual image ion optics with high transmission in a benchtop single-sector mass spectrometer (GCmateII from JEOL Ltd., Tokyo; Matsuda et al., 1974; Matsuda, 1976, 1981)

3.1.2 Quadrupole mass spectrometer

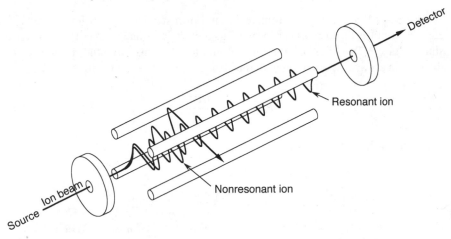

Fig. 3.4 Quadrupole mass spectrometer. The ion beam is accelerated to a high velocity by an electric field and passed through the quadrupole mass analyzer comprising four metal rods. DC and AC potentials are applied to the quadrupole rods in such a way that only ions with one mass-to-charge ratio (m/z) can pass though the analyzer at a time. To scan different m/z, DC and AC potentials are varied

3.1.3 Ion trap mass spectrometer

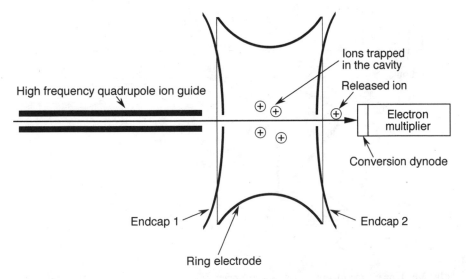

Fig. 3.5 Ion trap mass analyzer. With the help of different radio frequency signals applied to the ring electrode and the endcaps, all ions are trapped in the cavity and then sequentially ejected according to their m/z

3.1.4 Time-of-flight mass spectrometer

In time-of-flight mass spectrometers, a uniform starting time of ions is caused, e.g., by a pulse of an ionizing laser (Fig. 3.6) or a voltage pulse to an electric shutter (Fig. 3.7). After passing through the accelerating potential difference, V, the kinetic energy, E, of an ion with the charge, z, mass, m, and velocity, v, is:

$$E = zV = \frac{mv^2}{2}$$
(3.2)

For a length, l_{TOF}, the time of flight, t, is:

$$t = l_{TOF}\sqrt{\frac{m}{2zV}}$$
(3.3)

For example, for $l_{TOF} = 0.1$ m, $z = e = 1.602 \times 10^{-19}$ C, $m = 10$ kDa $= 10$ kg / 6.0221×10^{23}, and $V = 100$ V we obtain $t = 72$ μs. A mass resolution of 1 Da requires in this example a time resolution of 3.6 ns. Unfortunately, not all ions start to move at the same time and not all ions have the same velocity. The differ-

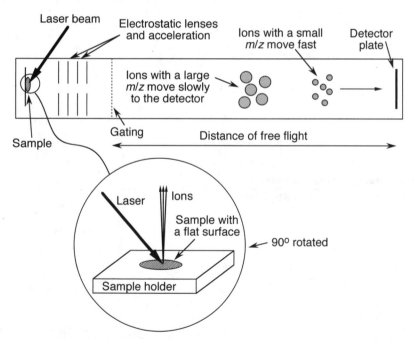

Fig. 3.6 Linear time-of-flight mass spectrometer (TOF) with matrix-assisted laser desorption ionization (MALDI). The linear configuration of TOFs represents the simplest implementation of the time-of-flight technique. The typical mass range lies between 0 and 100 kDa, and the typical mass resolution $m/\Delta m$ is 300–2000

ences in velocity are called chromatic aberration. Due to the chromatic aberration and the differences in the starting time, the requirements for a very high resolution are hard to meet in the simple design of a linear TOF (Fig. 3.6). In reflectron TOFs (Fig. 3.7–3.9), the ion optics reverses the flight direction of the ions and reduces chromatic aberration.

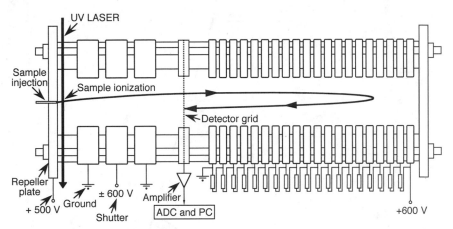

Fig. 3.7 A simple reflectron time-of-flight mass spectrometer (e.g., Bryden, 1995). The reflector enhances mass-spectrometric resolution: it increases the time of flight and can focus ions. Here a voltage pulse at the shutter electrode causes a uniform starting time of the ions

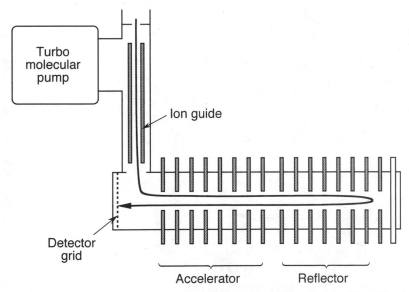

Fig. 3.8 A reflectron time-of-flight mass spectrometer with orthogonal ion inlet (e.g., BioTOF II from Bruker Daltonik, Bremen, Germany)

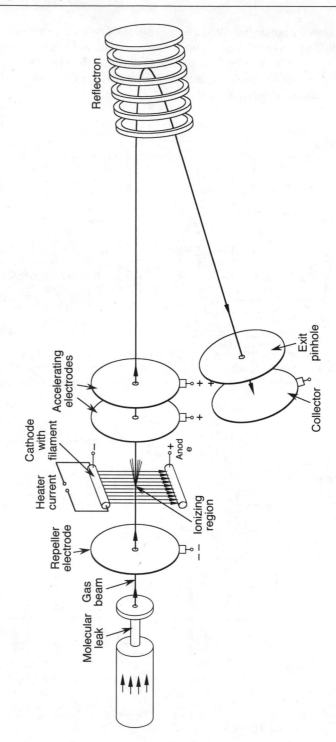

Fig. 3.9 A high-resolution reflectron time-of-flight mass spectrometer. For further details see, e.g., IonSpec Corp., Irvine, CA; JEOL USA, Inc., Peabody, MA; Micromass and Waters Corporation, Milford, MA; Thermo Finnigan, San Jose, CA; Varian Instruments, Walnut Creek, CA

3.1.5 Fourier transform mass spectrometer

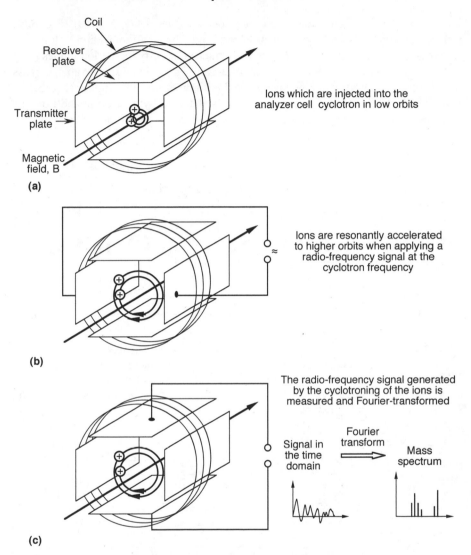

Coil

Receiver plate

Transmitter plate

Magnetic field, B

(a)

Ions which are injected into the analyzer cell cyclotron in low orbits

(b)

Ions are resonantly accelerated to higher orbits when applying a radio-frequency signal at the cyclotron frequency

The radio-frequency signal generated by the cyclotroning of the ions is measured and Fourier-transformed

Signal in the time domain

Fourier transform

Mass spectrum

(c)

Fig. 3.10 Principle of operation of a Fourier transform mass spectrometer (FTMS), also called "ion cyclotron resonance mass spectrometer" (see, e.g., IonSpec Corp., Lake Forest, CA; Bruker Daltonik, Bremen, Germany). (**a**) Ions are injected into the analyzer cell of the spectrometer. The magnetic field forces the thermal ions on orbits with small radii that depend on their mass-to-charge ratio. (**b**) An applied radio frequency pulse resonantly moves the ions to higher orbits. (**c**) The radio-frequency signal generated by the cyclotroning of the ions is measured and Fourier-transformed. For the method of Fourier transform see also Sect. 4.1.1. The striking characteristic of FTMS is the high resolution, R, typically in excess of 100,000

3.1.6 Ionization, ion transport and ion detection

Common methods of ionization are electrospray (Fig. 3.11), MALDI (Fig. 3.6), electron bombardment, ion bombardment, and chemical ionization. Ions are mainly guided by electrostatic lenses and quadrupole or octopole ion guides (Fig. 3.12). With the exception of FTMS, the ion signals emerging from the mass analyzer of the MS are commonly detected with an electron multiplier (Fig. 3.13). In FTMS the cyclotroning ions are indirectly detected by measuring and Fourier-transforming the voltage signal they induce into receiver electrodes.

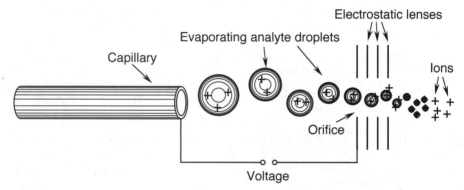

Fig. 3.11 Electrospray ionization method. Analyte solutions delivered by liquid chromatography or a syringe pump are sprayed through the narrow, heated capillary leading into the mass spectrometer. A voltage of typically 200 V – 5 kV is applied between capillary and orifice in front of the electrostatic lenses. Ions form in vacuum by evaporation of the analyte solution of charged droplets

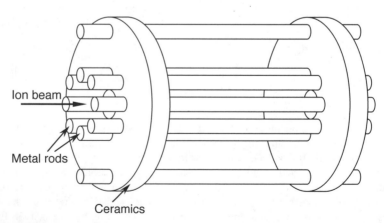

Fig. 3.12 High frequency octopole ion guide for the injection of ions into an ion trap MS. Compared with a quadrupole ion guide, it enables a higher precision of guidance

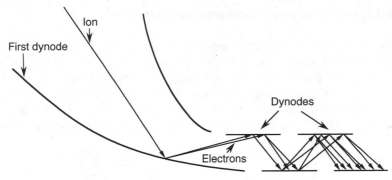

Fig. 3.13 Electron multiplier for a MS. The first dynode converts the ion current into an electron current. Further dynodes amplify the electrons by a total factor of typically 10^3–10^8, largely dependent on the electron accelerating voltage between the dynodes, the number of dynodes, and the dynode composition. The last dynode is connected with an ammeter (not shown)

3.1.7 Ion fragmentation

Significant enlargement of the information content of spectra is achieved by fragmenting the sample, e.g., in a collision chamber (Fig. 3.14) or a helium-containing cavity of an ion trap mass analyzer (Fig. 3.5; see also Sect. 3.2).

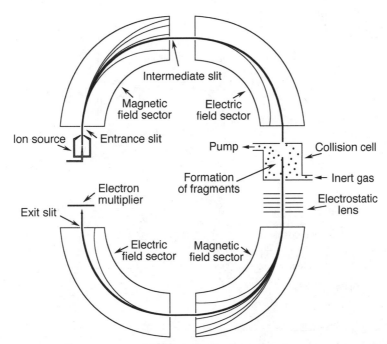

Fig. 3.14 High resolution sector MS with a collision chamber

3.1.8 Combination with chromatographic methods

For the study of highly complex systems, such as complete cells, MS is often combined with chromatographic methods, such as HPLC (high pressure liquid chromatography; Fig. 3.15), FPLC (fast performance liquid chromatography) and gas chromatography (GC; Figs. 3.16 and 3.17). The two types of connectors between chromatography and MS shown in Figs. 3.15 and 3.18 are both applicable for HPLC and FPLC. Two-dimensional spectra are obtained through

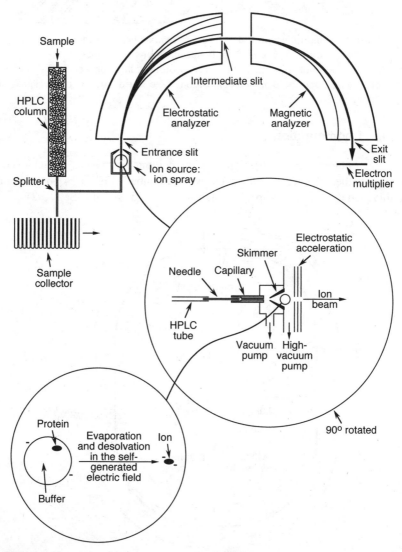

Fig. 3.15 Double sector MS in combination with HPLC

combination of mass spectrometry with chromatographic methods (e.g., Fig. 3.17). The resolution in two dimensions greatly enhances the analyzability of complex mixtures with a large number of components. For example, ion exchange chromatography on a crude cell extract with a resolution of 100 combined with mass spectrometry with a resolution of 10,000 can result in a total resolution of almost 1,000,000 for small and medium-sized soluble cellular proteins for which both methods are often largely independent from each other.

Buffer interference which is occasionally observed in MS can usually be prevented by increasing the sample concentration, decreasing the buffer concentration, or changing the buffer (Fig. 3.19).

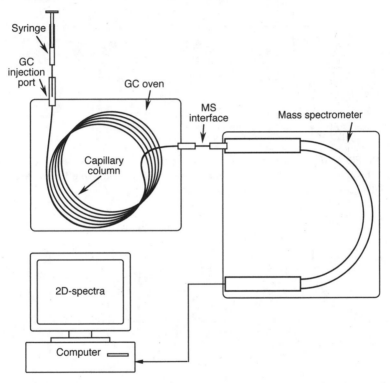

Fig. 3.16 GC/MS. The combination of mass spectrometry with gas chromatography can greatly enhance the resolution of complex samples

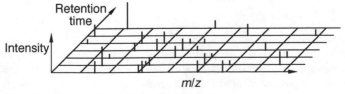

Fig. 3.17 Example of a two-dimensional GC/MS spectrum

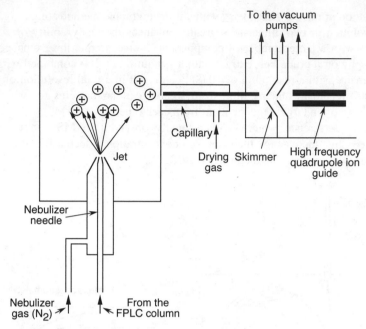

Fig. 3.18 FPLC/MS connector. In several stages the solvent is removed from the analyte solution by application of dry nitrogen and vacuum. The quadrupole ion guide leads the ions to the mass analyzer of the mass spectrometer

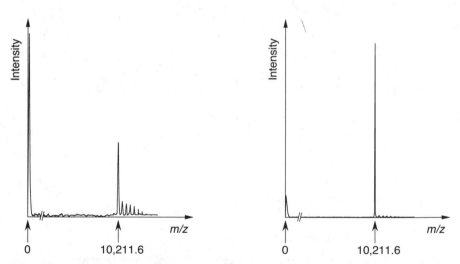

Fig. 3.19 Mass spectrogram of barstar, the 89-residue inhibitor of the ribonuclease barnase (Nölting et al., 1995, 1997a). *Left:* a number of side-peaks indicate the binding of buffer ions to the highly charged protein. *Right:* a measurement with a lower buffer concentration and higher protein concentration at a pH closer to the pI of the protein yields a cleaner mass spectrogram

3.2 Biophysical applications

A considerable interest in the fast point detection of toxic and non-toxic biological materials, such as certain bacterial strains, viruses, and proteins led to the development of portable mass-spectrometric biological detectors (e.g., Figs. 3.20 – 3.27;

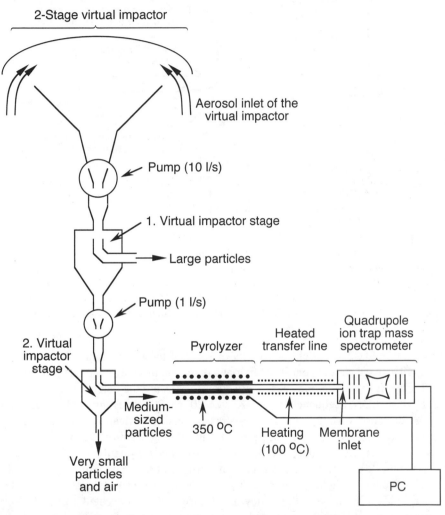

Fig. 3.20 Mass-spectrometric detector of biological agents, comprising a virtual impactor, a pyrolyzer, and a mass spectrometer: The two-stage virtual impactor selects particles of a certain size range, e.g., 1–10 μm (see also Fig. 3.21). These particles are then decomposed by pyrolysis (see Fig. 3.22) and analyzed by an ion trap mass spectrometer (see Figs. 3.24– 3.27). For a similar design see the CBMS (chemical-biological mass spectrometer) from Bruker Daltonik, Bremen, Germany

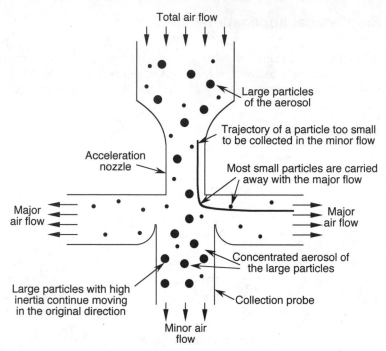

Total air flow

Large particles
of the aerosol

Trajectory of a particle too small
to be collected in the minor flow

Acceleration
nozzle

Most small particles are carried
away with the major flow

Major
air flow

Major
air flow

Concentrated aerosol of
the large particles

Large particles with high
inertia continue moving
in the original direction

Collection probe

Minor air
flow

Fig. 3.21 Principle of operation of a single-stage virtual impactor. The device splits the total flow of the aerosol into the minor and the major flow. Because the direction of the major air flow is perpendicular to the original direction of air flow, only particles with sizes smaller than the cut-off size can enter the major flow. In contrast, all large particles with a large inertia remain in almost the original direction of flow and join the minor flow containing also some of the small particles (Marple and Chein, 1980; Marple et al., 1998)

Williams et al., 2002). The sensitivity of some of these detectors is better than 1 biological agent particle per liter of air at a detection time of less than 3 minutes.

Further important biophysical applications of MS are the detection of mutations in DNA and DNA sequencing (e.g., Köster, 2001a, 2001b; Hung et al., 2002; Fig. 3.28), detection of mutations and post-translational modifications of recombinant proteins (e.g., Lee et al., 2002a; Sect. 9.1.2 in Nölting, 1999b; Fig. 3.29), diagnosis of diseases (e.g., Anderson et al., 2002), protein identification (e.g., Aitken and Learmonth, 2002; and Chap. 9), protein purity control (e.g., Stephenson et al., 2002), peptide sequencing (e.g., Shimonishi et al., 1981; Katakuse et al., 1982; Chen et al., 2002a; Nemeth-Cawley and Rouse, 2002; Shevchenko et al., 2002; Stoeva et al., 2002; Figs. 3.30 and 3.31), proteome analysis (e.g., Giffin et al., 2001; Nyman, 2001; Kersten et al., 2002; Lim et al., 2002; and Chap. 9), protein folding investigations (e.g., Birolo et al., 2002; Canet et al., 2002), protein conformational studies (e.g., Favier et al., 2002; Grandori et al., 2002), protein-protein interaction comparisons (e.g., Powell et al., 2002; Zal et al., 2002), and the search for extraterrestrial life (e.g., Fig. 3.32; Schwartz et al., 1995).

Pyrolysis tube

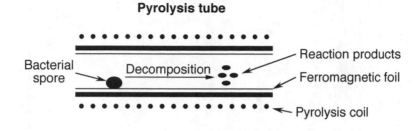

Temperature profiles

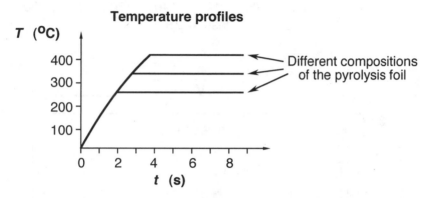

Fig. 3.22 Principle of operation of pyrolysis: The pyrolysis tube is loaded sequentially and a radio-frequency current passed through the pyrolysis coil. The current rapidly heats the ferromagnetic foil up to the Curie point where the foil reversibly ceases to exhibit ferromagnetic properties and further heating stops. Suitable ferromagnetic materials are, e.g., iron-nickel alloys. Pyrolysates generated in vacuum are then transferred to the mass spectrometer. The pyrolysis mass spectrometry (PyMS) method (Aries et al., 1986; Berkeley et al., 1990; Goodacre, 1994; Freeman et al., 1995; Goodacre and Kell, 1996) is particularly useful for the detection and analysis of biological agents

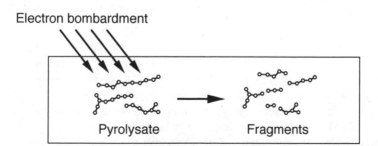

Fig. 3.23 Fragmentation of the pyrolysate by electron bombardment, typically in the energy region around 25 eV, further increases the information content concerning the nature of the biological agent under investigation (see, e.g., Ikarashi et al., 1991)

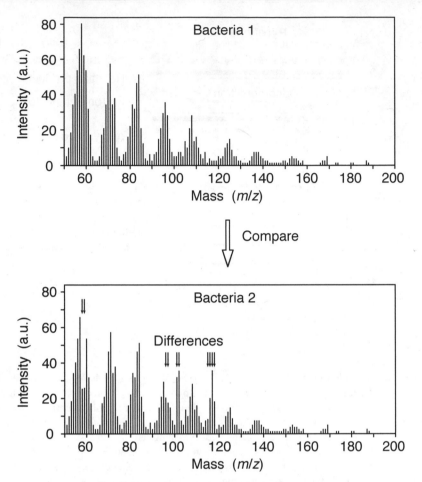

Fig. 3.24 Detection of biological contamination and differentiation between different biological agents by comparison of the mass spectra of the pyrolysate with m/z in the range 50–200 (see, e.g., Goodacre, 1994; Goodacre and Kell, 1996)

Especially for the analysis of highly complex biological systems, such as bacterial spores, the combination pyrolysis-MS (PyMS) is extraordinary useful. In this method the sample is partially decomposed in its components prior to mass-spectrometric analysis. The mass spectra of pyrolyzed biological systems may contain more than 100 lines, enabling a very sensitive differentiation of different samples. PyMS is used for the detection of bacteria, bacterial spores and viruses and the differentiation between different species of bacteria and viruses (Freeman et al., 1990, 1997; Snyder et al., 1990, 2001; Sisson et al., 1991; Sultana et al., 1995; Goodfellow et al., 1997; Helyer et al., 1997; Leaves et al., 1997; Magee et al., 1997; Timmins and Goodacre, 1997; Taylor et al., 1998; Barshick et al., 1999;

Goodacre et al., 2000; Tripathi et al., 2001), for the analysis of forensic samples and the personal identification of humans (Ishizawa and Misawa, 1990; Kintz et al., 1995; Armitage et al., 2001; Sato et al., 2001), and for biotechnological applications (Goodacre and Kell, 1996). PyMS spectra may be analyzed by using neuronal networks (Fig. 3.26; Goodacre et al., 1996, 1998a, 1998b; Nilsson et al., 1996; Kenyon et al., 1997). Ion trap mass spectrometers are particularly suitable for the pyrolysis-MS identification of biological agents since they can directly measure multiple fragmentation (Fig. 3.27).

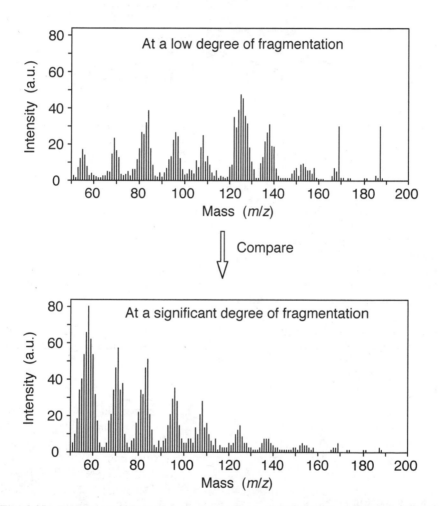

Fig. 3.25 Detection of biological contamination and differentiation between different biological agents by comparison the mass spectra of the pyrolysate with m/z in the range 50–200 at different degrees of fragmentation. The fragmentation is achieved, e.g., by collision with helium atoms in the cavity of an ion trap MS (see also Fig. 3.27)

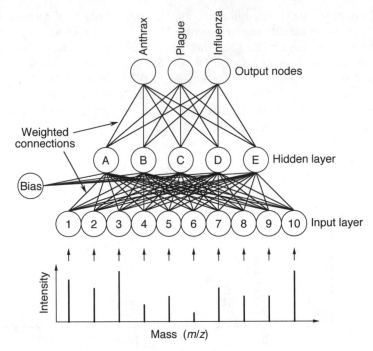

Fig. 3.26 Example of a neuronal network for the analysis of PyMS spectra (Kenyon et al., 1997; Goodacre et al., 1998a, 1998b). In this simplified example the network consists of only 10 input, 5 hidden, 3 output nodes, a bias, and weighted connections. The actual number of nodes in the PyMS input layer is usually equal to 150 (one for each m/z from 51 to 200). The hidden layer may actually contain 8–20 nodes. Preferentially the weights of the connections and the bias are set by supervised learning using the hazardous substances to be detected or simulants of these hazardous substances

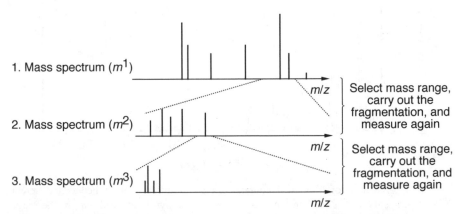

Fig. 3.27 Example of mass spectrometry with multiple fragmentation (m^3) in an ion trap mass spectrometer

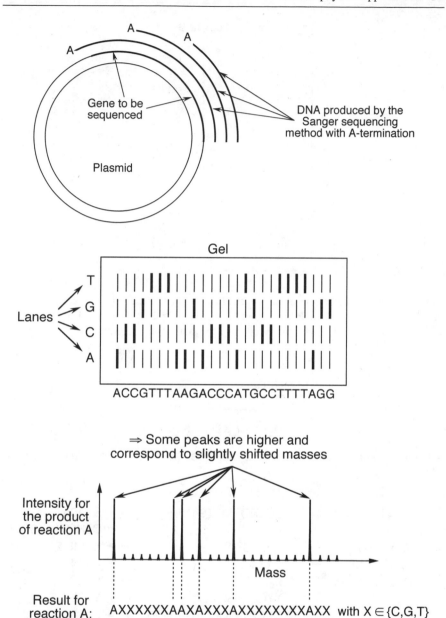

Fig. 3.28 DNA sequencing with the Sanger dideoxynucleotide termination method (Sanger et al., 1977; Sanger, 1988) and mass-spectrometric detection. *Top:* amplification of the template DNA with termination (only one of the four reactions is shown). *Middle:* for comparison to the mass-spectrometric method: conventional analysis of the reaction products with polyacrylamide gel electrophoresis. *Bottom:* analysis of the products of reaction A by mass spectrometry: the pattern of mass peaks shows the positions of adenine (A) in the sequence

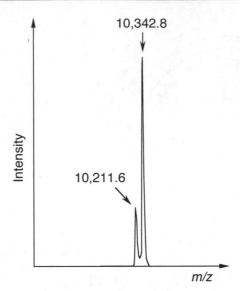

Fig. 3.29 Mass spectrogram of the 10,212-Da protein barstar. This protein preparation contains a fraction with a molecular weight 131 Da higher than expected. This is due to an N-terminal methionine which is not properly cleaved after protein synthesis

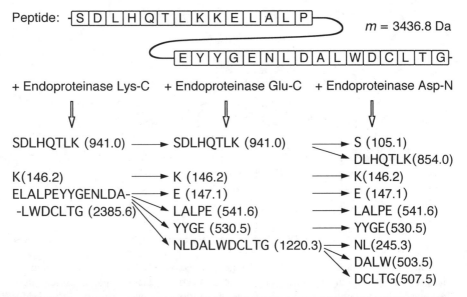

Fig. 3.30 Sequencing of the peptide SDLHQTLKKELALPEYYGENLDALWDCLTG by proteolytic digestion and mass spectrometry. This peptide corresponds to the helix$_1$-helix$_2$ peptide of the protein barstar (Nölting et al., 1997a). In this example with only three proteases, only some parts of the peptide sequence can unambiguously be identified

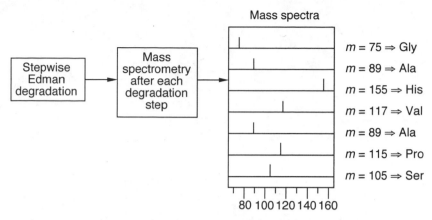

Fig. 3.31 Peptide sequencing using Edman degradation and mass spectrometry (Shimonishi et al., 1981; Katakuse et al., 1982): reaction products of sequential degradation are mass-spectrometrically identified

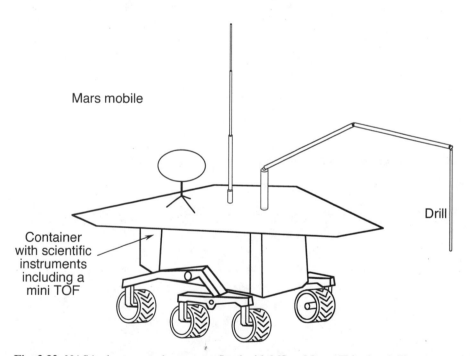

Fig. 3.32 NASA plans to send rovers outfitted with MS to Mars. This sketch illustrates a mobile with mini-TOF searching for extraterrestrial life. Already now mass-spectrometry is utilized to search for extraterrestrial bacteria in meteors

4 X-ray structural analysis

4.1 Fourier transform and X-ray crystallography

X-ray crystallography is the method with the highest currently available resolution power for structures of large macromolecules and macromolecular complexes. Since the technique of Fourier transform is central to this method, we first deal with some essential aspects of this technique:

4.1.1 Fourier transform

Mathematically the Fourier transform and inverse Fourier transform convert between two domains (spaces), e.g., the domain r (e.g., space or time) and the domain k (e.g., momentum or frequency):

Fourier transform: $$F(k) = (2\pi)^{-0.5} \int_{-\infty}^{\infty} f(r) \, e^{-2\pi i r k} \, dr \qquad (4.1)$$

Inverse Fourier transform: $$f(r) = (2\pi)^{-0.5} \int_{-\infty}^{\infty} F(k) \, e^{2\pi i r k} \, dk \qquad (4.2)$$

Fig. 4.1 illustrates a one-dimensional Fourier transform: (b) Represents the decomposition of the signal from (a). This decomposition was calculated from the Fourier transform (c). The inverse transform of (c) yields back exactly (a). (d) Is the inverse Fourier transform of the signal of (c) with all phases set to zero instead of using the correct phases. The comparison of (d) with (a) illustrates the importance of the phases in Fourier transform: in order to be able to correctly obtain back the original signal by inverse Fourier transform, both the amplitudes and the phases have to be known.

Three examples in Figs. 4.2–4.4 demonstrate the method of two-dimensional Fourier transform: Fig. 4.2b represents the Fourier transform of the hexagonal arrangement of peaks of Fig. 4.2a. In Fig. 4.3a some of Fourier components with low amplitudes are set to zero, and yet the inverse Fourier transform (Fig. 4.3b) shows that essentially all information is still preserved. Fourier transforming a noisy object, then substituting certain low-amplitude parts of the Fourier transform by zeros, and then inverse-transforming the modified Fourier transform, is an efficient method for noise reduction. Figs. 4.4a and 4.4b show the result when using only a small slice of the Fourier transform for the calculation of the inverse

Fourier transform: significant distortions are observed along the coordinate for which too few Fourier components were utilized for reconstruction.

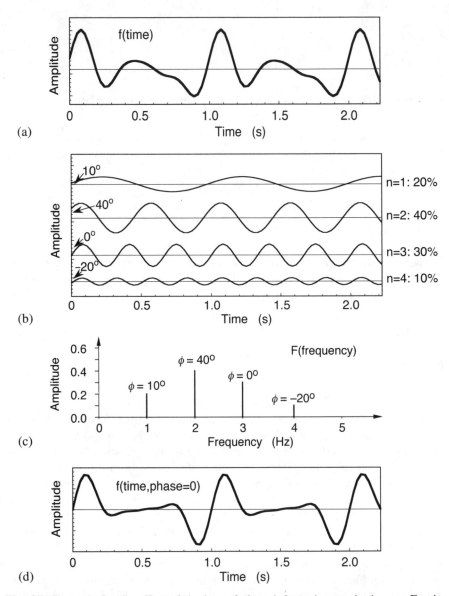

Fig. 4.1 Example for the effect of the loss of phase information on the inverse Fourier transform: With the correct phases, the four frequency components (**b**) add to the sum shown in (**a**). However, when adding the four components with the wrong phase 0, we obtain the wrong sum (**d**). (**c**) Represents the Fourier transform of (**a**). In (**a**), (**b**), (**c**) only a fraction of the function is shown; the complete function is periodical in $(-\infty, \infty)$

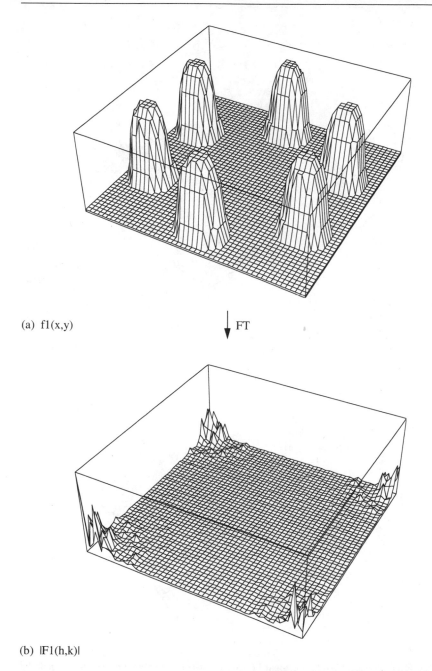

(a) f1(x,y)   ↓ FT

(b) |F1(h,k)|

Fig. 4.2 Example of a two-dimensional Fourier transform: (**b**) is the Fourier transform of (**a**). Only the absolute of the function is drawn in (**b**). However we must keep in mind that the complete function contains an amplitude and phase for each coordinate point (compare with Fig. 4.1)

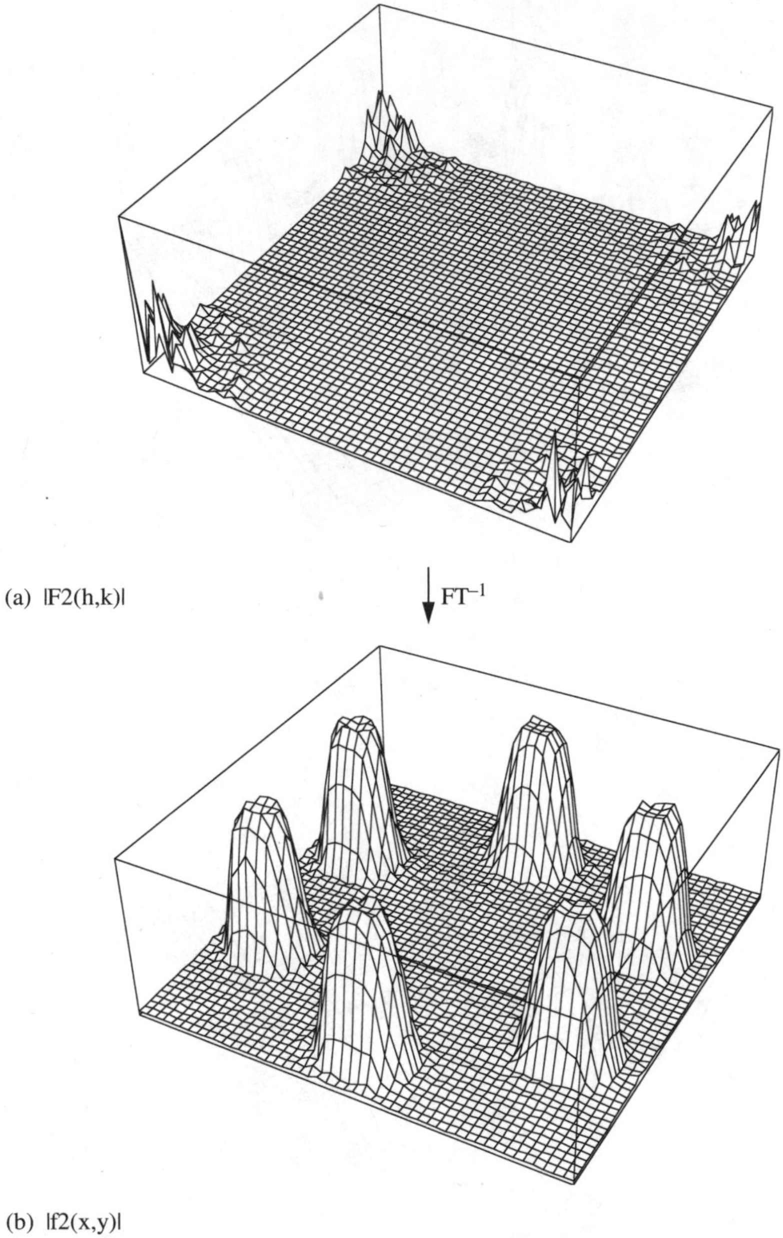

(a) |F2(h,k)| ↓ FT⁻¹

(b) |f2(x,y)|

Fig. 4.3 Example of a two-dimensional Fourier transform: In the corners, (**a**) is identical to Fig. 4.2b, but has the components with low amplitude in the middle of the coordinate space set to zero. (**b**) Is the inverse Fourier transform of (**a**). (**b**) Is found to be almost identical to Fig. 4.2a showing that the deleted low-amplitude components did not contain much information. Note that in this figure only the absolutes of the functions are drawn

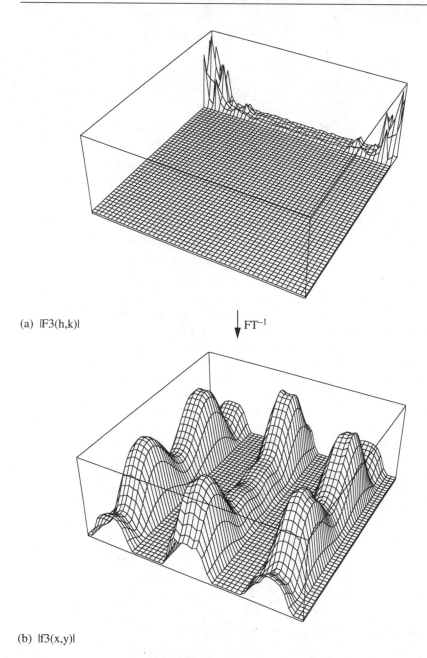

(a) |F3(h,k)|

FT^{-1}

(b) |f3(x,y)|

Fig. 4.4 Example of a two-dimensional Fourier transform: (a) Is a thin slice of the function in Fig. 4.2b plus a thick slice with all components set to zero. (b) Represents the inverse Fourier transform of (a): some features of the function in Fig. 4.2a are still preserved in (b), but much information is lost. Note that in this figure only the absolutes of the functions are drawn

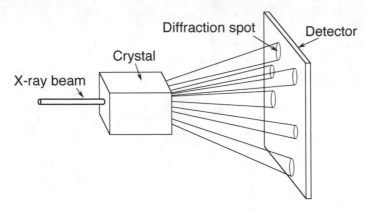

Fig. 4.5 Example of a diffraction experiment on a crystal. The X-ray diffraction pattern of the crystal is recorded with an area detector. The pattern consists of a large number of discrete spots

Why is the Fourier transform so important for X-ray crystallography? This is because the diffraction pattern of a crystal (Fig. 4.5) or any other physical object is the Fourier transform of its structure (see also later Fig. 4.10).

To understand why the diffraction pattern of a physical object is its Fourier transform let us consider the diffraction of a wave by a single object (Fig. 4.6) and two point-sized objects separated by r (Fig. 4.7): the scattering vector, S, is defined as $S = s/\lambda - s_0/\lambda$, where s, s_0, and λ, are the vector of the incident wave, vector of the diffracted wave, and wavelength, respectively (Fig. 4.6). Then the phase difference in units of wavelengths between the two waves in Fig. 4.7 is given by: $rs/\lambda - rs_0/\lambda = rS$. Constructive interference of the two waves occurs in case of $rS = 0, \pm 1, \pm 2, \ldots$; destructive interference, i.e. extinction, is observed at $rS = \pm 1/2, \pm 3/2, \ldots$. The diffraction pattern, $\mathbf{F}(S)$ of the two points is then given by $\mathbf{F}(S) = e^{-2\pi i r S}$, with i being the imaginary number defined as $i \equiv \sqrt{-1}$.

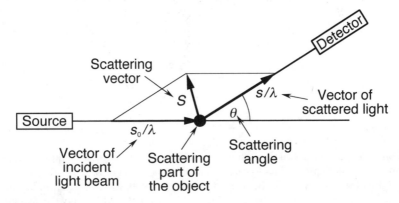

Fig. 4.6 Diffraction of a wave by a single part of an object

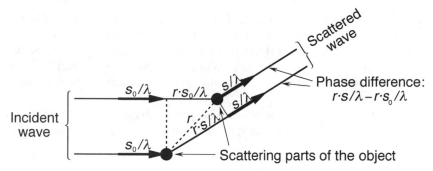

Fig. 4.7 Diffraction of a wave by a two objects of equal scattering power

For a the diffraction, $F(S)$, of a macroscopic object consisting of many diffracting points with varying diffraction power, $\rho(r)$, we have to integrate all scattered waves:

$$F(S) = \int_{-\infty}^{\infty} \rho(r)\ e^{-2\pi irS}\ dr \tag{4.3}$$

This equation has exactly the form of a Fourier transform (compare with Eq. 4.1). Hence the electron density and structure of a protein can be obtained from the inverse Fourier transform of its diffraction image.

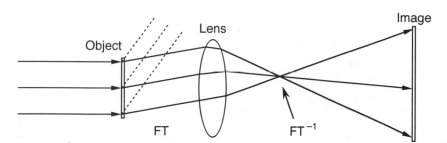

Fig. 4.8 A lens projecting the image of an object onto a screen performs an inverse Fourier transform of the diffraction pattern of the object

In microscopes the inverse Fourier transform is performed by lenses (Fig. 4.8). Unfortunately, currently there is no X-ray microscope with sufficient resolution and sensitivity. X-ray mirrors do not provide sufficient resolution, and because of radiation damage, we would not obtain a satisfactory resolution for a single protein molecule anyway. That is why we have to record the diffraction pattern of a protein crystal and to calculate the inverse Fourier transform of the diffraction pattern with a computer. Unfortunately, when recording the diffraction pattern of an object with the help of a camera, all phase information is lost. With other words, we do not record the complete Fourier transform, but only a fraction of it. The consequences of this serious problem were illustrated in Fig. 4.1. Thus, additionally to the recording of the diffraction pattern, one needs a special technique to recover the phase information. The currently most important method to recover phase information in protein crystallography on new structures is the technique of heavy atom replacement (see Sect. 4.1.2.4).

A specifics of the diffraction of macroscopic crystals is that not a continuous diffraction pattern is obtained, but discrete spots. To understand this behavior, consider the structure of a crystal (Fig. 4.9):

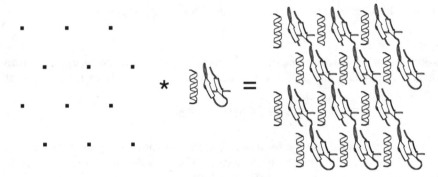

Fig. 4.9 Mathematically a protein crystal can be described as the convolution of the crystal lattice with the unit cell (one or a few protein molecules)

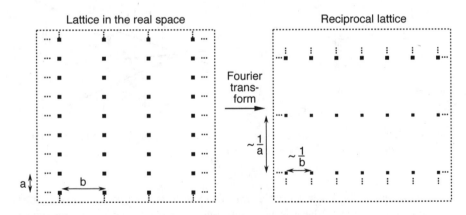

Fig. 4.10 The Fourier transform of the crystal lattice is the so-called reciprocal lattice. It determines the maximum number and positions of the observed diffraction spots

The protein crystal can be described as the convolution of the crystal lattice with the unit cell (Fig. 4.9): crystal = lattice * unit cell. The unit cell is the smallest unit from which the crystal can be generated by translations alone. It usually contains one or several protein molecules. According to the convolution theorem, the Fourier transform, FT, of two convoluted functions $f_1(r)$ and $f_2(r)$ is the product of their Fourier transforms:

$$FT (f_1(r) * f_2(r)) = FT (f_1(r)) \cdot FT (f_2(r)) \qquad (4.4)$$

Thus, the diffraction pattern of a protein crystal is the Fourier transform of the unit cell times the Fourier transform of the crystal lattice. The latter is called reciprocal lattice (Fig. 4.10). Since the reciprocal lattice is zero outside its lattice points, the crystal diffraction pattern corresponds to the Fourier transform of the unit cell sampled at the points of the reciprocal lattice.

A second way to explain the occurrence of discrete spots in the diffraction pattern of macroscopic crystals, and to evaluate the information from the intensity of theses spots, is to think of the diffraction as a reflection on the X-ray at the lattice planes of the crystal (Fig. 4.11). These lattice planes are described by the Miller indices (Fig. 4.12).

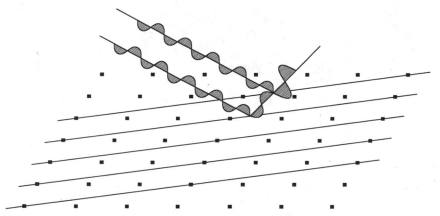

Fig. 4.11 Reflection of X-rays at the lattice planes of a crystal. Diffraction is viewed as reflection of the X-ray on the lattice planes

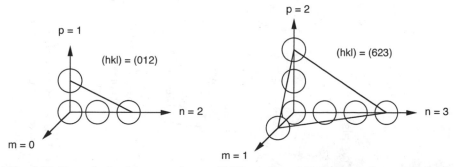

Fig. 4.12 Example for the nomenclature of Miller indices, hkl. Miller indices are defined as the smallest integer multiple of the reciprocal axis sections in which 1/0 is set to 0

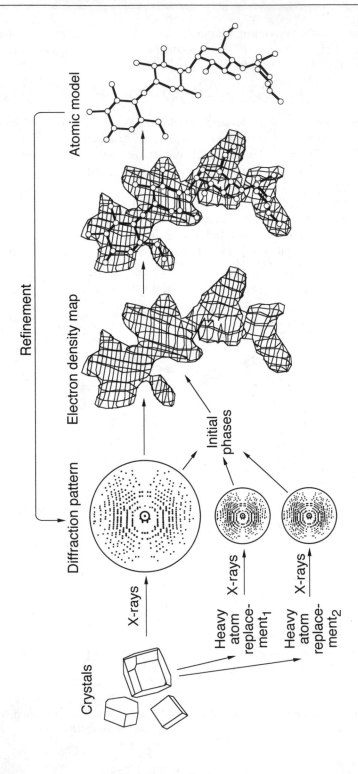

Fig. 4.13 Overview of X-ray crystallographic analysis of proteins: From the measured diffraction pattern of suitable native and, if necessary, heavy atom replaced crystals, an initial electron density and atomic model is calculated. The initial model is refined, e.g, by modifying it till its calculated diffraction pattern matches the measured pattern.

4.1.2 Protein X-ray crystallography

4.1.2.1 Overview

In 1934 Bernal and Crowfoot discovered that pepsin crystals give a well-resolved X-ray diffraction pattern (Bernal and Crowfoot, 1934; Bernal, 1939). It took 3 decades and the development of computers to obtain the first 3-D structures of proteins (Kendrew et al., 1960; Perutz et al., 1960). Many thousands of native protein structures have been solved since then. Examples are found in Figs. 1.6–1.8. A few structure determinations were even made under artificial conditions, e.g., in organic co-solvents (Schmitke et al., 1997, 1998).

4.1.2.2 Production of suitable crystals

For X-ray diffraction we must have a single crystal of suitable geometry and size (Fig. 4.13 on the previous page and Figs. 4.14–4.16). Commercial crystal screening kits, containing the most prominent buffers for protein crystallization, may be obtained, e.g., from JenaBioScience (Jena, Germany). Important parameters for coarse-screening and fine-adjustment are protein concentration, salt types and concentrations, pH, type and concentration of surfactants and other additions, temperature, and speed of crystallization.

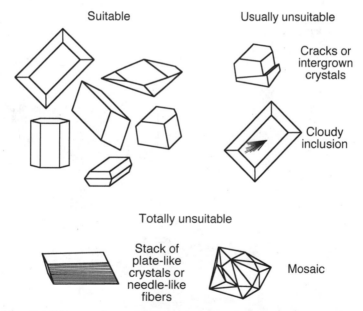

Fig. 4.14   Suitable protein and virus crystals are transparent and do not have inhomogeneities of color or refractive index. Crystals with cracks, intergrown crystals and crystals with cloudy inclusions are generally unsuitable for X-ray crystallography. Totally unsuitable are stacks of plate-like crystals or needle-like fibers and mosaics

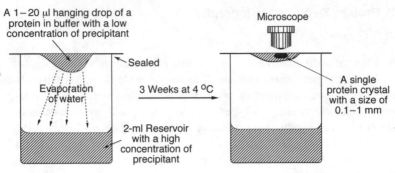

Fig. 4.15 Hanging drop method. The solvent of a small drop of protein or virus solution attached to a cover slide slowly evaporates partially. At the right conditions, a single crystal of suitable size grows

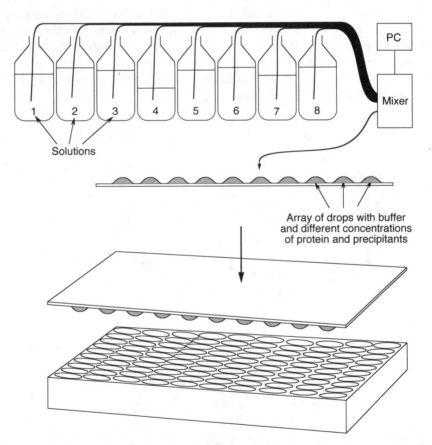

Fig. 4.16 Crystallization robot for hanging drop crystallization. The computer-controlled mixer draws different solutions from reservoir bottles, mixes them with various ratios, and places the mixtures on a glass plate

One generally starts with a protein concentration of about $2-50$ mg ml^{-1}. Usually, the protein or virus must not contain a significant amount of contaminants, such as other proteins or viruses, protein or virus fragments, unfolded or misfolded protein, particulate matter, chemical additions unnecessary for stability or solubility. In most cases compact proteins that do not contain floppy ends, such as histidine tags or native unstructured peptides, crystallize better. Suitable crystals have sizes of a few 0.1 mm.

4.1.2.3 Acquisition of the diffraction pattern

For the acquisition of the crystal diffraction pattern (Figs. 4.17–4.19), multi wire area detectors or CCD area detectors (Fig. 4.20) are commonly used. With the example of a linear CCD, Fig. 4.21 illustrates the basic principle of operation of CCDs.

The most common X-ray sources for protein and virus crystallographic analysis are rotating anode generators (Fig. 4.22) with typically $5-25$ kW electrical power and synchrotrons (Figs. 4.23 and 4.24). Synchrotrons are comparably expensive, but have a higher brightness enabling shorter measuring times. Reduction of the exposition time often results in a better quality of the diffraction pattern since decomposition of the crystal due to radiation damage is reduced.

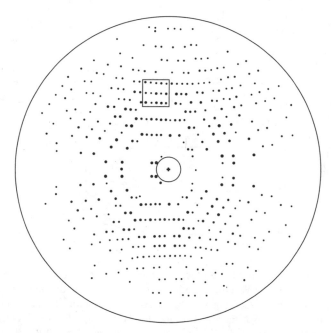

Fig. 4.17 X-ray diffraction pattern of a protein crystal (Norledge et al., 1996). The highlighted section is referred to in Fig. 4.26

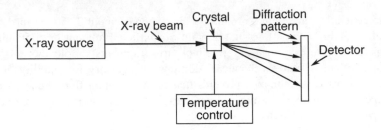

Fig. 4.18 General setup for the acquisition of the diffraction pattern

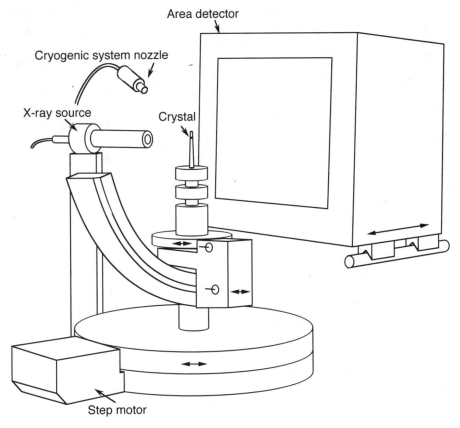

Fig. 4.19 Setup for acquiring the diffraction pattern with an area detector (see, e.g., area detectors from Rigaku, The Woodlands, TX). The crystal is cooled with nitrogen from the cryogenic system nozzle. Cooling the crystal reduces radiation damage, but somewhat changes the intermolecular distances. Diffraction of the X-rays from the X-ray source by the crystal are recorded with the area detector with typically 2048 × 2048 or 4096 × 4096 pixels (see next figure)

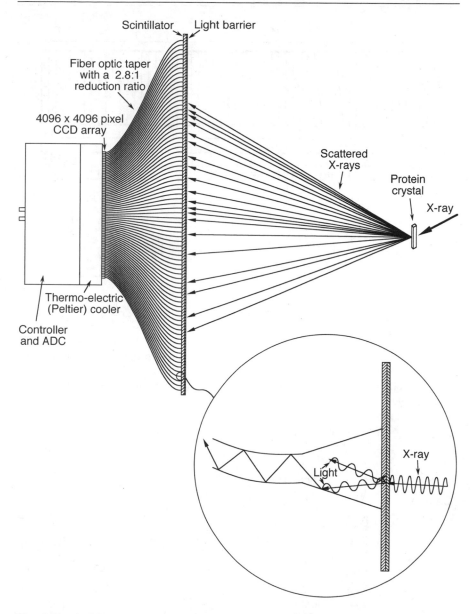

Fig. 4.20 A CCD area detector used for recording of X-ray diffraction patterns. For reduction of the dark current, this CCD is operated at −40 °C, allowing it to detect single photons. The fiber optic taper serves also for blockage of X-rays and thereby prevention of radiation damage to the sensitive CCD array. At a pixel size of 20 μm × 20 μm, the full well capacity is typically several 100,000 electrons per pixel, enabling the necessary high dynamic range

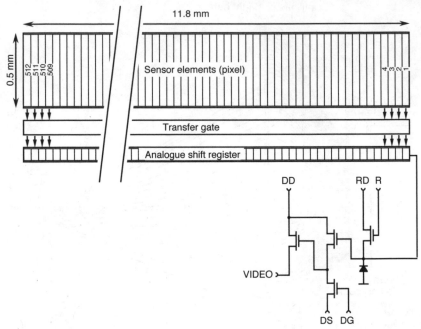

Fig. 4.21 Linear charge coupled device (CCD). The sensor elements generate electrons by absorption of photons and store the electrons in potential wells. After a certain period of time, the collected electrons are transferred to the analogue shift register and read out (Nölting, 1991)

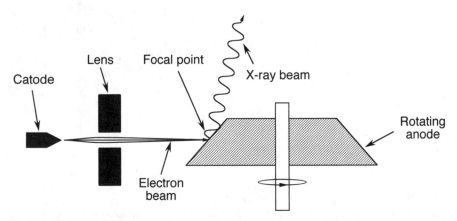

Fig. 4.22 Rotating anode generator. An electron beam is focussed onto the rotating anode. It knocks out electrons from the inner electron shells of the anode metal. Reoccupation of the vacant shells by electrons from higher level shells involves the emission of X-ray radiation. The interaction of the electron beam with the anode metal generates also a large amount of heat which is quickly dissipated by rotating the anode below the spot of incidence of electrons

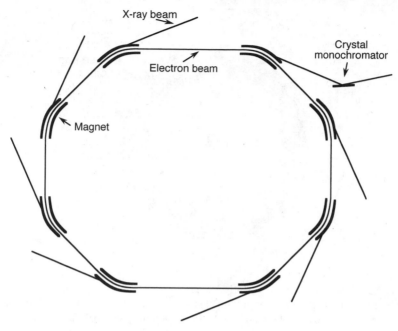

Fig. 4.23 Design of a synchrotron. Ions or electrons are accelerated to a speed close to the speed of light and forced on a curved trajectory. A broad spectrum of radiation is produced along the curved sections of the beam. For protein crystallography, a certain wavelength, e.g., 1 Å, is selected by a monochromator

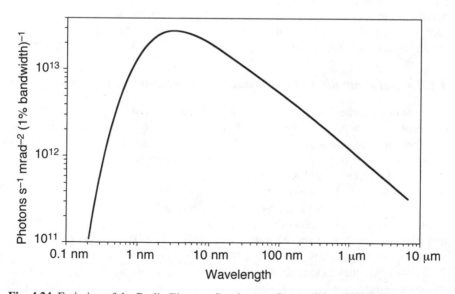

Fig. 4.24 Emission of the Berlin Electron Synchrotron Storage Ring (BESSY I)

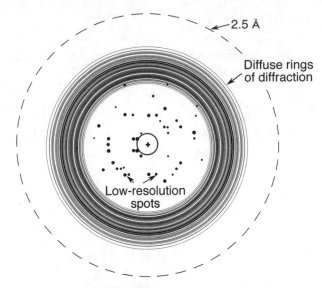

Fig. 4.25 Diffraction pattern of a poorly scattering crystal: only a few spots near the center are observed, any high resolution information is absent. The dashed circle indicates the area corresponding to a resolution of 2.5 Å. If diffraction spots would be visible up to this circle, the resolution of the obtained structure would be 2.5 Å. One can see that the resolution is much lower in this example

Already superficial inspection of the diffraction pattern provides a lot of information about the quality of the crystals: since the information about fine details of the protein structure is found at large diffraction angles, the absence of spots far outside the center of the diffraction pattern shows that only a low resolution will be obtained (Fig. 4.25).

4.1.2.4 Determination of the phases: heavy atom replacement

As mentioned earlier, after measurement of the diffraction pattern, determination of the phase information is required. If we do not have information from molecules with a similar structure, or anomalously scattering atoms in the molecule, the method of choice may be the heavy atom replacement: the diffraction pattern of the original (native) crystal is compared with crystals that contain a single or a few heavy atoms at fixed positions. Those crystals can be prepared, e.g., by diffusing a solution of a heavy atom salt into the protein crystal.

Fig. 4.26 depicts sections of the diffraction pattern of the native protein crystal and heavy atom derivatized crystal, respectively. The diffraction spots labeled with "++" are significantly increased in intensity for the heavy atom derivative. This shows that they belong to phases with a large magnitude. How can we make this conclusion? See Fig. 4.27 which, in the upper part, shows the intensity of an interferogram of two waves as function of phase: when we introduce a small shift to one of the waves (lower part), a large increase of intensity of the interferogram is found for large positive phases. Essentially no change of intensity occurs at phases around zero. Thus, analogously we can conclude that diffraction spots which increase in intensity only slightly between native crystal and heavy atom derivative belong to phases around zero or π. So, by comparing the intensities of the spots between native crystal and the heavy atom derivative we can estimate the phases of the individual diffraction spots. With only one heavy atom derivative, an uncertainty of two possibilities remains for each spot, but this can easily be removed with a further, different heavy atom derivative of the protein crystal.

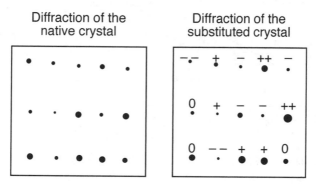

Fig. 4.26 Section of the diffraction pattern of a protein crystal. *Left:* "native" crystal. *Right:* heavy atom derivative

Another way of showing the importance and meaning of phases in crystallography is illustrated in Fig. 4.28: Atoms with different phases and relative positions may cause a diffraction spot at the same position. Thus, without information from heavy atom replacement, or from diffraction patterns of proteins with similar structure or other information, we cannot deduce the protein structure from the diffraction pattern. Theoretically one could also try out all possible phases and see if it leads to a meaningful structure, but currently for macro-molecules the computational effort would be much too high.

It should be noted that the problem of loss of phase information occurs only in the common methods of recording the crystal diffraction, such as with a photographic film or a semiconductor detector. The use of lenses or mirrors to produce an image like in an microscope would prevent this loss of information (see p.65). Unfortunately, currently we cannot build a lens which is sufficiently suitable for focussing X-rays of less than a few Å wavelength: the surface of a conventional lens would not be smooth enough and the bulk of the lens would act like a non-regular grating. Further, it is also very difficult to build highly precise X-ray mirrors (Figs. 4.29 and 4.30). X-ray mirror microscopes using soft radiation currently reach only a few 10 nm resolution. More importantly, the radiation damage would prevent atomic resolution of a single protein molecule or virus.

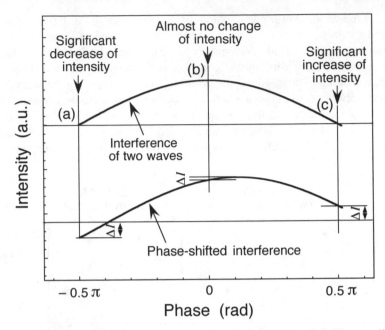

Fig. 4.27 Interferogram of two waves: within the phase interval $[-\pi/2, \pi/2]$, a small phase shift causes a large negative amplitude change, ΔI, for large negative phases (a), and a large positive amplitude change for large positive phases (c), but almost no amplitude change for zero phases (b). Thus, e.g., from a large amplitude increase of a diffraction spot upon application of a small phase shift by an additional heavy atom, we can conclude that the phase of the spot has a large magnitude. Analogously one can estimate the phases from the observation of various intensity changes of diffraction spots upon derivatization of the crystal with a heavy atom. For the complete phase interval, $(-\pi, \pi]$, there are still two phases for each amplitude change (not shown). This uncertainty is removed by using data from a second heavy atom derivative

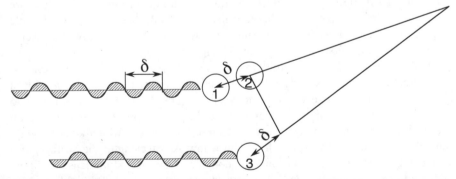

Fig. 4.28 Different phases and relative positions may cause a diffraction spot at the same position: the atom pairs (1,2) and (2,3) with completely different relative locations cause positive interference at the same position

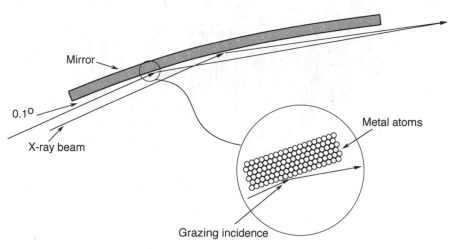

Fig. 4.29 Common X-ray mirrors are only suitable for low angles of incidence

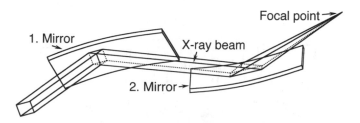

Fig. 4.30 A pair of X-ray mirrors with grazing incidence focuses an X-ray beam to a spot

The mathematics behind the method of heavy atom replacement is exemplary illustrated in Figs. 4.31–4.34. Fig. 4.31a represents an array of atoms. Fig. 4.31b is the absolute of the Fourier transform of this array. From the imaginary and real parts (Fig. 4.32) of this Fourier transform, the phase was calculated (Fig. 4.33). Fig. 4.34a represents the same array as above, but with one additional heavy atom causing a small change of the Fourier transform. The difference of the absolutes of Fourier transforms between native array and heavy atom derivatized array is shown in Fig. 4.34b. Comparing this difference of the absolutes of Fourier transforms with the absolutes of the phases of the native array (Fig. 4.33b), we find a connection between phases (Fig. 4.33b) and amplitude differences (Fig. 4.34b). This connection allows the magnitude of the phase angles to be determined. As mentioned, the remaining ambiguity of sign is removed by including the data from a second isomorphous heavy atom derivative.

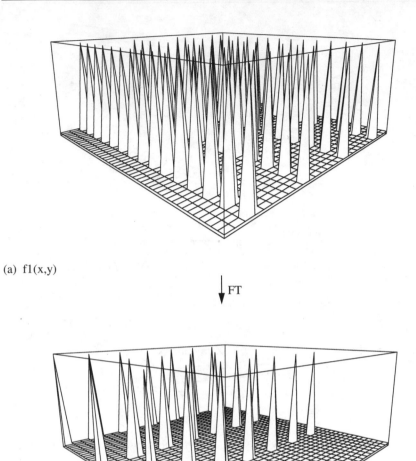

(a) f1(x,y)

$\downarrow$ FT

(b) |F1(h,k)|

Fig. 4.31 (a) Representation of an array of atoms. (b) Absolute of the Fourier transform of (a)

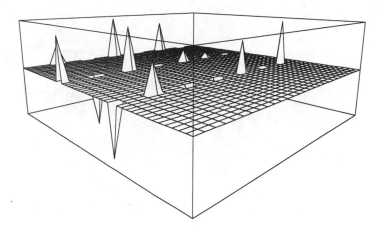

(a) Im(F1(h,k))

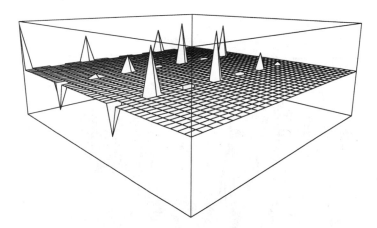

(b) Re(F1(h,k))

Fig. 4.32 (**a**) Imaginary part of the Fourier transform of the array of Fig. 4.31a. (**b**) Real part of the Fourier transform of the array of Fig. 4.31a

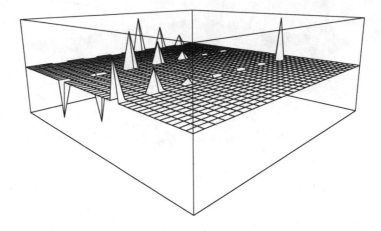

(a) Phase(F1(h,k))

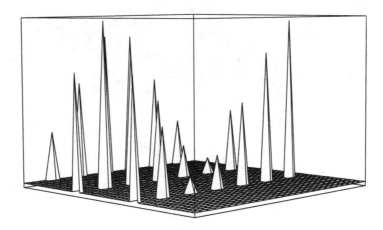

(b) |Phase(F1(h,k))|

Fig. 4.33 Phase (**a**) and absolute of the phase (**b**) of the Fourier transform of the array of Fig. 4.31a, calculated from imaginary and real parts of the Fourier transform

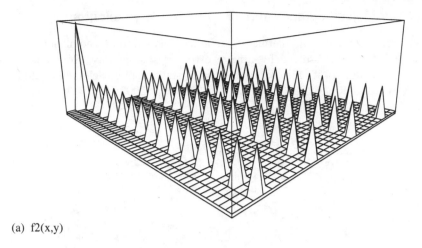

(a) f2(x,y)

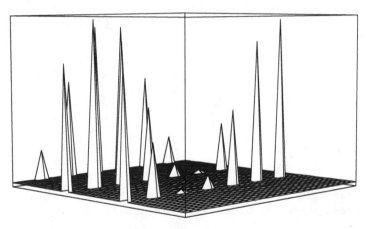

(b) |F1(h,k)| – |F2(h,k)|

Fig. 4.34 (**a**) Representation of the array from Fig. 4.31 with an additional heavy atom. (**b**) Difference of the absolutes of the Fourier transforms of native array (Fig. 4.31a) and heavy atom replaced array (Fig. 4.34a). When we compare Figs. 4.33b and 4.34b, we can see a connection between phases and amplitude differences. Note that $|F| = ((\text{Im}(F))^2 + (\text{Re}(F))^2)^{0.5}$; $\text{phase}(F) = \arctan(\text{Im}(F)/\text{Re}(F))$, where "Im" and "Re" stand for imaginary and real parts, respectively

4.1.2.5 Calculation of the electron density and refinement

Software for the calculation of the initial electron density from the diffraction data and the refinement of structures is being rapidly developed by several academic institutions and often supplied for free. It may be found on the internet, e.g., by searching with the keywords "protein crystallography software".

4.1.2.6 Cryocrystallography and time-resolved crystallography

Short-living conformational intermediates in the microsecond and nanosecond time scale have been resolved by time-resolved crystallography (Srajer et al., 1996; Genick et al., 1997; Fig. 4.35) and cryocrystallography (Schlichting et al., 2000; Petsko and Ringe, 2000; Wilmot and Pearson, 2002; Fig. 4.36). Time-

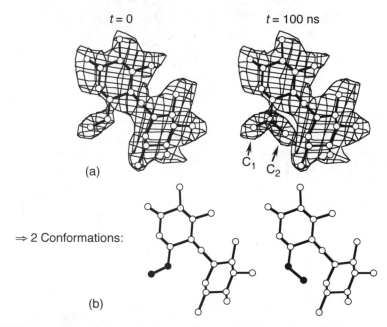

Fig. 4.35 Example for time-resolved crystallography. 100 ns after initiation of a conformational change, the electron density indicates the occurrence of two conformations, C_1 and C_2

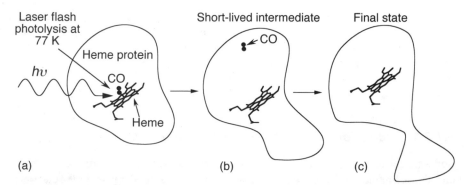

Fig. 4.36 Example for cryocrystallography. The CO is flashed off the heme group of the heme protein. This initiates a conformational transition which is detected, e.g., at –196 °C

resolved crystallography interprets time-dependent electron density maps and can offer detailed structural information on short-lived intermediates under near-physiological conditions. In cryocrystallography, reactions are induced and measured at a low temperature. At the very low temperatures of flash photolysis and acquisition of the diffraction pattern in the experiment shown in Fig. 4.36, the reaction kinetics of the conformational changes is slowed down by many orders of magnitude. This enables to determine the coordinates of structural intermediates that would normally be too short-lived to be resolved by X-ray crystallography.

4.2 X-ray scattering

4.2.1 Small angle X-ray scattering (SAXS)

Small angle X-ray scattering serves for the elucidation of microstructural information in amorphous materials on length scales ranging from a few Å to a few μm (Figs. 4.37 and 4.38). Fig. 4.39 is an illustration of the setup for SAXS. Significant effort is undertaken to enable the measurement at very small angles. Since there is an about reciprocal relationship between distance separation of scattering points (Δx) and the scattering angle (θ), this measurement is essential to obtain sufficient information in the relatively large length scale compared with the wavelength (λ) of the X-rays ($\Delta x \approx 0.5\lambda \sin^{-1}(\theta/2)$). For details on X-ray optics see also the previous section.

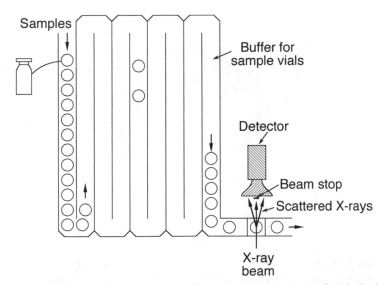

Fig. 4.37 Automatic SAXS and wide angle X-ray scattering analysis of biological samples. A large number of vials is automatically sampled and production faults immediately are detected and responded to

SAXS measurements revealed that

(a) an unliganded aspartate transcarbamoylase adopts a T-quaternary structure (Fetler et al., 2002),

(b) the axial period of collagen fibrils is 65.0 ± 0.1 nm in healthy human breast regions, and 0.3 nm larger in cancer-invaded regions (Fernandez et al., 2002),

(c) flax cellulose microfibrils probably have a cross section of 10×50 Å^2 (Astley and Donald, 2001),

(d) microfibrils with an axial repeating period of approximately 8 nm are present in the major ampullate silk from the spider *Nephila* (Miller et al., 1999; Riekel and Vollrath, 2001), and

(e) the ATPase domain of SecA has dimensions of approximately 13.5 nm $\times$ 9.0 nm $\times$ 6.5 nm (Dempsey et al., 2002).

SAXS revealed information regarding the conformational diversity and size distribution of unfolded protein molecules (Kamatari et al., 1999; Panick et al., 1999a; Garcia et al., 2001; Choy et al., 2002), and was used in a large number of protein-folding and peptide-folding studies to obtain information about size changes (e.g., Chen et al., 1998; Panick et al., 1998, 1999b; Arai and Hirai, 1999; Segel et al., 1999; Kojima et al., 2000; Russell et al., 2000; Aitio et al., 2001; Canady et al., 2001; Katou et al., 2001; Muroga, 2001; Tcherkasskaya and Uversky, 2001). SAXS is one of the very few methods which can directly monitor structural changes of small virus particles (Sano et al., 1999; Perez et al., 2000).

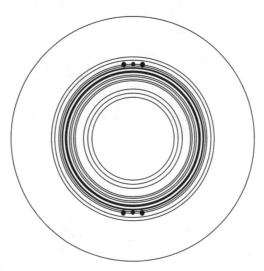

Fig. 4.38 Diffraction pattern of a cell suspension. SAXS can serve to obtain a "fingerprint" of a biological specimen which helps to identify unknown biological samples

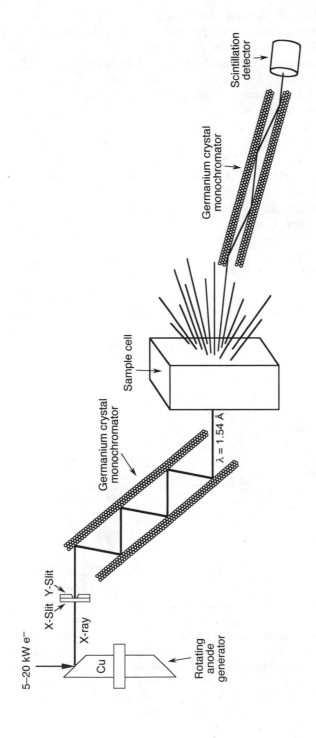

Fig. 4.39 Setup for small angle X-ray scattering (SAXS). The X-ray beam from a rotating anode generator is passed through a crystal monochromator that selects a wavelength. The monochromatic X-ray beam is passed though the sample cell, and scattered X-rays at very small angels are passed through a second crystal monochromator and then detected with a scintillation detector. For many applications, this set-up may be simplified, e.g., by choosing fewer reflections in the monochromators which yields a higher intensity of X-rays

Distance constraints derived from SAXS measurements can be used to filter candidate protein structures for the purpose of protein structure prediction (Zheng and Doniach, 2002). In some cases even low resolution solution structures of proteins were obtained solely from SAXS data (Chacon et al., 1998; Shilton et al., 1998; Bada et al., 2000; Maruyama et al., 2001; Scott et al., 2002) or SAXS data combined with neutron scattering data (Egea et al., 2001). SAXS can reveal the structure of bones (Rinnerthaler et al., 1999) and structural changes in bones due to diseases (Grabner et al., 2001). The method was used to obtain information about conformational changes of bacterial cell wall enzymes upon binding to a substrate (Schönbrunn et al., 1998), and structural changes in artificial biological membranes (Riske et al., 2001). SAXS results on human dentin, which is a complex composite of collagen fibers and carbonate-rich apatite mineral phase, are consistent with nucleation and growth of an apatite phase within periodic gaps in the collagen fibers (Kinney et al., 2001).

4.2.2 X-ray backscattering

The property of X-rays to penetrate materials is used in many biophysical applications, ranging from for the purpose of determination of the molecular weight of

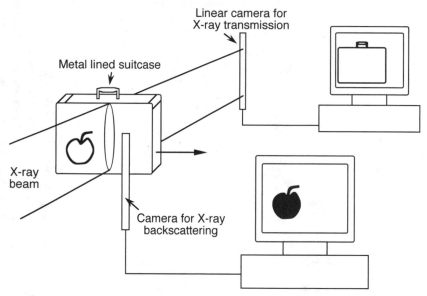

Fig. 4.40 Detection of biological and other organic material behind a metal layer with X-ray backscattering: since the absorption of biological material is much smaller than that of metal, the biological material is difficult to detected in single-wavelength X-ray absorption measurements. X-ray backscattering provides much better contrast in this application. However, the quantum efficiency of X-ray scattering is low and thus relatively large expositions and sensitive cameras must be used

proteins to X-ray backscattering for the purpose of detection of organic material hidden in metal containers (see, e.g., Fig. 4.40).

A problem of detection of organic material, such as illicit drugs and explosives, by X-ray absorption is their low absorption coefficient compared with metals and the possibility to camouflage the material, e.g., by embedding it in other organic material, such as flour or sugar. X-ray backscattering offers a good contrast for the detection of such powdery material (Fig. 4.40). The main disadvantage is the low backscattering coefficient compared with transmission coefficient of most organic samples. Thus, a significantly higher exposure compared with X-ray transmission is usually required.

5 Protein infrared spectroscopy

Infrared spectroscopy is based on the infrared absorption of molecules and is, compared with crystallography, a relatively simple and inexpensive tool for the global characterization of molecular conformations and conformational changes of proteins and other biomolecules. Depending on the measurement technique, scanning infrared (IR) spectrometers, Fourier transform infrared (FTIR) spectrometers, and single wavelength infrared apparatuses are distinguished (see Sect. 5.1). Typically the most interesting spectral region for biomolecules is $v = 400$ cm^{-1}– 4000 cm^{-1}, where the wavenumber, v, is defined as $v \equiv 1/$wavelength. Infrared activity requires a change of dipole moment upon excitation (Fig. 5.1). For proteins the amide chromophore absorption in the region of 1500 cm^{-1}–1700 cm^{-1} ($\approx 6\,\mu$m wavelength) is particularly important for the assessment of secondary structure content and structural changes. Regarding the resolution of protein secondary structure, the information content of IR and FTIR spectroscopy is comparable with that of circular dichroism (see, e.g., Nölting et al., 1997b; Nölting, 1999b), and regarding the resolution of features of the tertiary structure of proteins, IR and FTIR are often inferior, and yet IR is much easier to apply on a fast time scale and for remote sensing (see, e.g., LIDAR in Sect. 5.1.3).

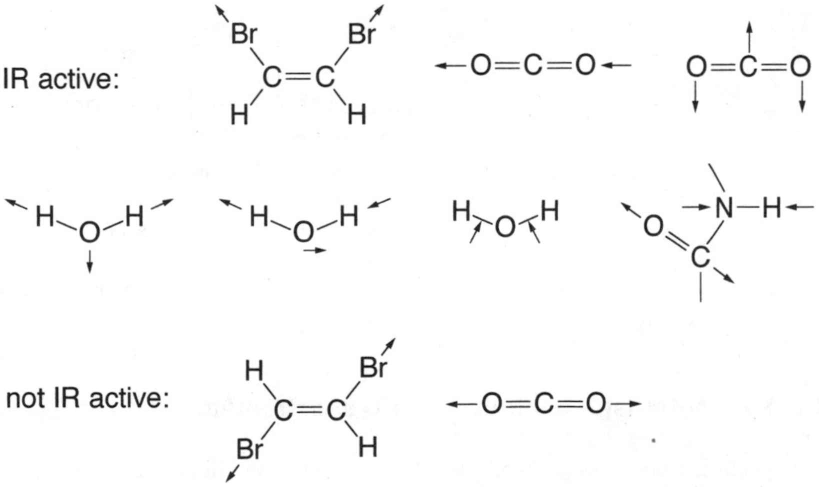

Fig. 5.1 Example of infra-red active and non-active vibrations. Note that infra-red activity requires a change of dipole moment

5.1 Spectrometers and devices

5.1.1 Scanning infrared spectrometers

Early IR spectrometers (Fig. 5.2) were constructed similarly to scanning UV/VIS absorption spectrometers. The emission of the source, e.g., a thermal source operated at 1000 °C, is passed through a monochromator selecting a single wavelength. The monochromatic beam is split into two beams – one having the sample in the path. A shutter passes through only one of the two beams at a time. Both beams are alternatingly detected by an IR detector, e.g., a pyroelectric detector, and compared which each other. The optical density of the sample is calculated from the logarithm of the intensity quotient. The use of light modulation is quite indispensable since the problem of background radiation is much more severe than in UV/VIS spectrometers. Spectra are recorded by scanning the wavelength region of interest. This scanning principle of operation is still widely used in IR spectrometers with time resolutions in the femtosecond to nanosecond region, where infrared lasers serve as IR source (see Nölting, 1999b).

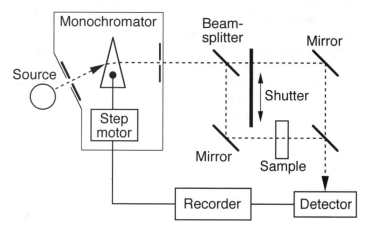

Fig. 5.2 Example of a scanning infrared (IR) spectrometer. The monochromator separates the radiation of the IR source into its different wavelengths and selects one wavelength at a time. A beam splitter separates the monochromatic beam into sample beam and reference beam. The absorption coefficient, according to the chemical and structural properties of the sample molecules, is calculated using the detected intensity quotient between both beams, the pathlength, and the sample concentration

5.1.2 Fourier transform infrared (FTIR) spectrometers

FTIR spectrometers (Figs. 5.3–5.7) use the technique of Michelson interferometry and have the advantage of using a larger part of the emission of the IR source during the measurement of a spectrum, compared with scanning IR spectrometers that are based on monochromators which select only one wavelength at a time. The better usage of radiation improves the inherent signal-to-noise ratio, especially for strongly absorbing samples for which the measurement may be photon shot noise limited. Also the spectral resolution of FTIR spectrometers, which is limited by the path length of the moving mirror, is often better than that of scanning IR spectrometers.

In FTIR spectrometers (Fig. 5.3) the beam of radiation from the IR source is focused on a beam splitter constructed such that half the beam is transmitted to a moving mirror and the other half is reflected to a fixed mirror. Both the moving mirror and the fixed mirror reflect the beam back to the beam splitter which reflects the half of both beams to the detector where they interfere according to their phase difference. The light intensity variation with optical path difference, called interferogram, is the Fourier transform of the incident light spectrum (light intensity as a function of the wavenumber). Absorption spectra are obtained by measuring interferograms with a sample and with an empty sample cell in the beam and inverse Fourier transforming the interferograms into spectra (Figs. 5.4–5.6).

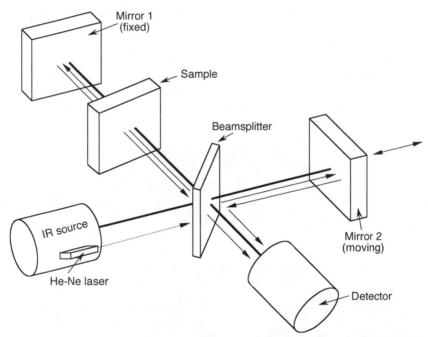

Fig. 5.3 Typical design of FTIR spectrometers. The lamp, e.g., a thermal source, emits a beam of infrared radiation. A Michelson interferometer, consisting of a beamsplitter, a fixed mirror and a moving mirror, splits the beam into two beams and generates an interference of them. The sample inserted in one of the beam paths changes the interference. Interferograms with and without sample are recorded and the absorption of the sample is calculated by inverse Fourier transform (see Fig. 5.6)

Fig. 5.4 shows three examples of interference of the two monochromatic light beams of the interferometer resulting in different intensities of the interferogram. Eq. 5.1 describes the intensity of the interferogram, $I_{\text{interferogram}}$, for the interference of two polychromatic beams of equal intensity in the FTIR spectrometer:

$$I_{\text{interferogram}}(\delta) = const \times \int_0^\infty I_{\text{beam}}(v)\cos(2\pi\delta v)\,dv \,, \tag{5.1}$$

where δ is the phase difference of the two beams, $const$ a constant, I_{beam} the intensity of the beams, and v the wavenumber. From the interferogram, the intensity of the beams can be calculated by inverse Fourier transform:

$$I_{\text{beam}}(v) = const \times \int_{-\infty}^{+\infty} I_{\text{interferogram}}(\delta)\cos(2\pi\delta v)\,d\delta \tag{5.2}$$

Analogously, the intensity of the beam with the sample in the path is calculated from the corresponding interferogram. The absorption is given by the logarithm of the intensity quotient of blank to sample.

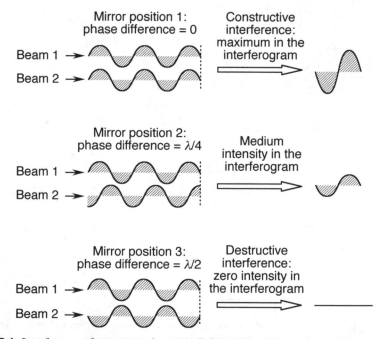

Fig. 5.4 Interference of two monochromatic light waves with equal intensity. *Top:* both beams have the same phase; their interference yields the maximum of the interferogram, i.e. the sum of both intensities. *Middle:* at a phase difference of $\lambda/4$, the intensity of the interferogram equals the intensity of the interfering beams. *Bottom:* at a phase difference of $\lambda/2$, both beams extinguish each other

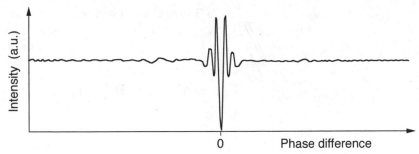

Fig. 5.5 Example of an interferogram of two polychromatic light beams

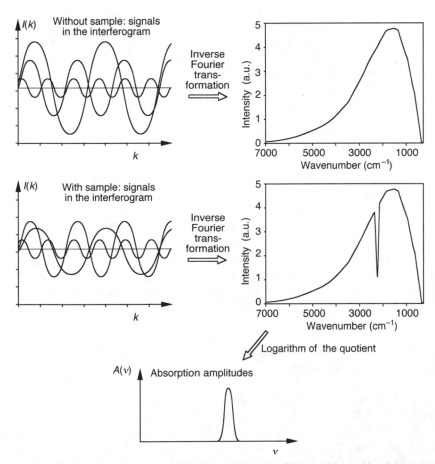

Fig. 5.6 Principle of operation of a FTIR spectrometer. IR intensities at the detector are recorded both for the sample cell filled with solvent and for the sample cell filled with sample. Inverse Fourier transform of the two interferograms yields the IR intensities. The IR absorption spectrum is calculated using the logarithm of the intensity quotient

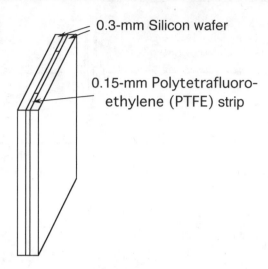

0.3-mm Silicon wafer

0.15-mm Polytetrafluoro-
ethylene (PTFE) strip

Fig. 5.7 Sample cell for FTIR experiments. The transparent walls of the cell are made from silicon wafers supplied by a manufacturer of electronic chips

A very suitable material for the manufacture of sample cells, sample holders, and windows is silicon (Fig. 5.7). Polished silicon wafers of 0.5 – 1 mm thickness are sufficiently transparent from 400 to 4000 cm^{-1} (25 – 2.5 µm wavelength) (Jiang et al., 1996). Only the fragility and the high refractive index of this material might be problematical in some experimental set-ups. Used infrared sources are often thermal sources operated at about 1000 °C. Beam splitters made from a thin germanium film evaporated on a potassium bromide (KBr) or cesium iodide (CsI) slide are transparent down to about 400 cm^{-1} (25 µm wavelength) and 200 cm^{-1} (50 µm wavelength), respectively. Liquid nitrogen cooled mercury cadmium telluride (MCT) detectors and deuterated triglycine sulfate (DTGS) pyroelectric detectors are frequently applied for infrared detection. For an excellent introduction into the instrumentation of FTIR spectroscopy see Perkins, 1986.

5.1.3 LIDAR, optical coherence tomography, attenuated total reflection and IR microscopes

IR spectroscopy is exquisitely suitable for remote sensing of clouds of biological agents (Fig. 5.8). The IR LIDAR set-up consists of a pulsed IR laser and an IR detector which senses the backscattered light from the laser. Since the light travels extremely fast, the detector senses the return echo before the next pulse is sent. The time it takes for the laser pulse to travel down and back is a measure of the distance. Mobile commercial LIDAR systems quite often employ an integrated global positioning system (GPS) to determine the own position.

Equipped, e.g., with an optical modulator which rapidly changes the direction of the beam, and mounted on top of a roof, the IR LIDAR can scan the 360°-environment at distances of 0 to several 10 km. This method has importance, for example, for early warning systems of smog in large cities and for three-dimensional analysis of forest structure and terrain (Fig. 5.9). Remote sensing of changes in forest structure utilizes the information of time and intensity of multiple reflections from leaves and branches. Effects of environmental pollutants and pests are quickly detectable in vast areas and economic damage is largely reducible.

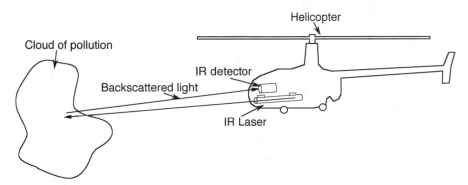

Fig. 5.8 Remote sensing of environmental changes, e.g., a cloud of biological material, with an IR LIDAR (light detection and ranging; measurement of light backscatter)

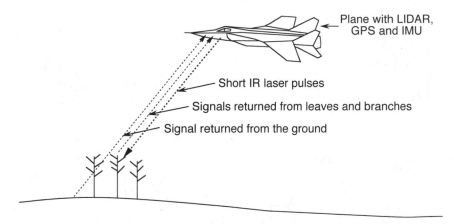

Fig. 5.9 Remote survey of forest structure and terrain with IR LIDAR technology. The plane is equipped with a GPS and an inertial measurement unit (IMU). The latter contains several gyroscopes and an accelerometer and can determine the position and angle of tilt with some accuracy during periods of failure of the GPS

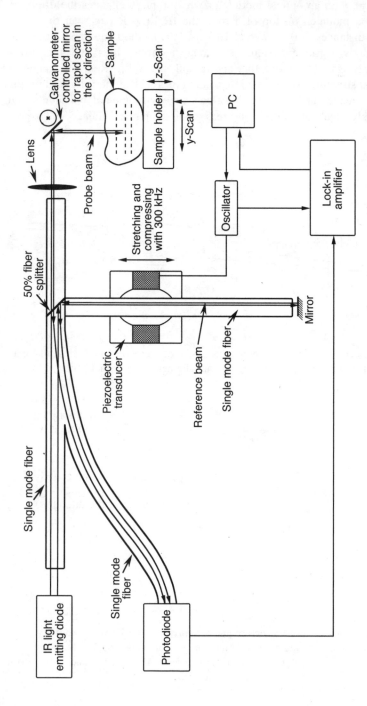

Fig. 5.10 Optical coherence tomography (OCT). The IR light from a light emitting diode is split into reference and probe beams. Light of the probe beam reflected from the sample is interfered with light of the reference beam, and the interference is detected by the photodiode. The pathlength of the reference beam is modulated by stretching an optical fiber with a piezoelectric transducer. Light from the sample which has traveled the same distance as the reference beam interferes constructively. Its signal is extracted from the interference intensity by a lock-in amplifier (Duncan et al., 1998)

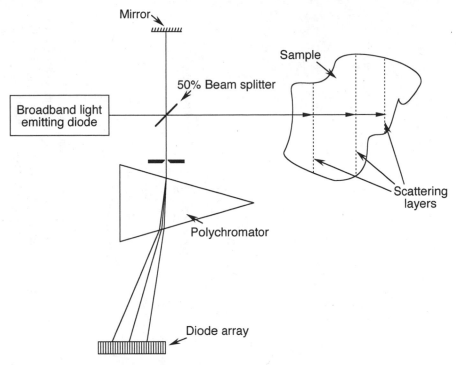

Fig. 5.11 Spectral domain optical coherence tomography (SDOCT) (Andretzky et al., 1998; Häusler and Lindner, 1998). Polychromatic backscattered light from different depths interferes with polychromatic light of a reference beam. The interference of the beams is analyzed with a polychromator and a multichannel detector. From the spectral changes due to interference, information about the depth of the scattering layer is obtained

Another important variant of IR spectroscopy on biological samples is optical coherence tomography (OCT). OCT (Figs. 5.10 and 5.11) utilizes echoes of infrared light waves backscattered off the internal microstructures within biological objects to obtain images on a μm scale. In the design of Fig. 5.10, IR radiation backscattered from the sample is interfered with a reference beam. Light from a scattering layer in the sample with a certain depth has the same phase as the reference beam and thus interferes constructively, i.e. produces a high interference intensity. Light from slightly deeper or shallower scattering layers cause a lower interference intensity. By modulating the phase of the reference beam and detecting the interference intensity with a lock-in amplifier, the signals from layers with different depths are extracted from the interference intensity (Duncan et al., 1998). Fig. 5.11 depicts a second design variant of optical coherence tomography (Andretzky et al., 1998; Häusler and Lindner, 1998). Here the information on depth is gained by analyzing the spectrum of the backscattered light.

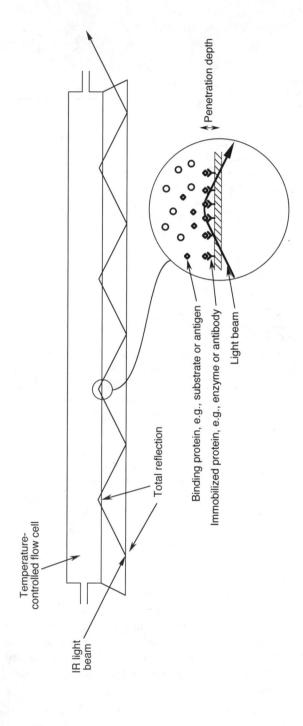

Fig. 5.12 Flow cell for attenuated total reflection (ATR) infrared spectroscopy (Fringeli et al., 1998; Snabe and Petersen, 2002). The internally total-reflected light slightly leaves the waveguide and so can probe the sample molecules on the outside of the waveguide. The part of the light wave which leaves the waveguide at the total reflection points is called evanescent wave. Only very little sample is needed. Using a large number of reflections can lead to a more than 100-fold amplification of the measured signal

The next IR spectroscopic technique to be mentioned is attenuated total reflection (ATR) infrared spectroscopy (see, e.g., Fringeli et al., 1998; Ding et al., 2002; Feughelman et al., 2002; Snabe and Petersen, 2002; Figs. 5.12 and 5.13). Here the coefficient of internal total reflection of an IR beam in a waveguide is changed by a sample deposited on the surface of the waveguide. An advantage of ATR on thin layered samples is the dramatic increase of the effective optical path-length and sensitivity through multiple reflections compared with conventional transmission spectroscopy on such a sample.

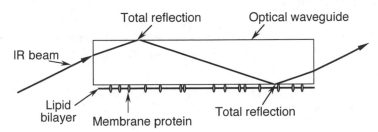

Fig. 5.13 Attenuated total reflection (ATR) infrared spectroscopy on membrane proteins (see, e.g., Ding et al., 2002)

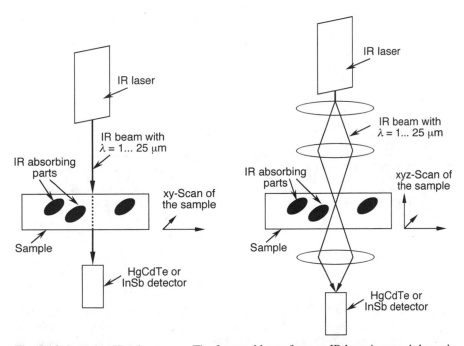

Fig. 5.14 Scanning IR microscope. The focussed beam from an IR laser is passed through the sample and detected. *Left:* simple microscope with planar resolution, especially suitable for thin layer samples. *Right:* microscope with 3-dimensional resolution: for acquisition of the image, the sample is moved in xyz-directions

Because of the significantly lower scattering of IR light relative to light of shorter wavelength, IR microscopes (Fig. 5.14) enable the inspection of most strongly scattering samples. Computer aided image processing allows two- or three-dimensional resolution. More complicated microscopes may utilize step-scan interferometry for photoacoustic depth profiling, monochromators for spectral analysis and polarizers/analyzers for linear dichroism (LD) analysis.

5.2 Applications

One of the biophysical main applications of FTIR is the characterization of the structure and conformational changes of proteins (Barrera et al., 2002; Butler et

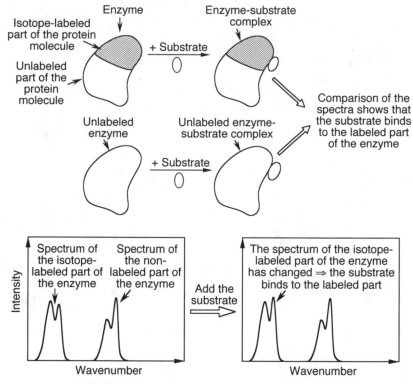

Fig. 5.15 Isotope-edited FTIR spectroscopy (see, e.g., Li et al., 2002; Barth, 2002). Since the spectrum of the isotope-labeled part of the protein molecule is significantly shifted, it can be distinguished from the spectrum of the non-labeled part. A change of the protein IR spectrum upon binding of the substrate to the protein shows which part of the molecule the substrate binds to. In this example, the magnitude of a peak in the spectrum of the isotope-labeled part of the protein has changed upon binding of the substrate. This shows that the substrate binds to the labeled part of the enzyme

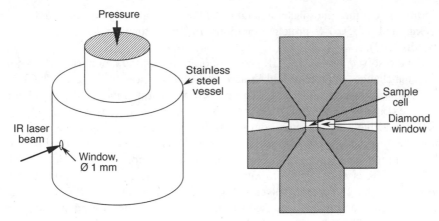

Fig. 5.16 Apparatus to monitor protein unfolding under high pressure with IR. Since the volume of unfolded protein is less than that of folded protein, high pressure favors transition to the unfolded state

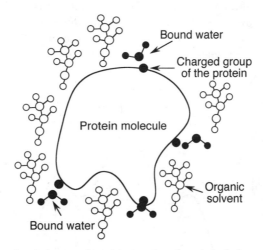

Fig. 5.17 Protein molecule in organic solvent: only a few strongly bound water molecules remain attached to the protein molecule

al., 2002; Castellanos et al., 2002; Dong et al., 2002; Hilario et al., 2002; Moritz et al., 2002; Mui et al., 2002; Noinville et al., 2002), of peptides (Bianco et al., 2002; Gordon et al., 2002; Huang et al., 2002; Torres et al., 2002), and of DNA (Lindqvist and Graslund, 2001; Malins et al., 2002). In some cases, interactions were resolved at the level of individual amino acid residues (Kandori et al., 2002; Mezzetti et al., 2002; Zhang et al., 2002a).

Isotope-edited FTIR is particularly useful for the structural characterization of specific macromolecular regions (Fig. 5.15): e.g., the three phosphate stretching vibrations of the phosphate calcium ATPase complex were detected at a background of 50,000 protein vibrations in an isotope exchange experiment (Barth, 2002).

Time-resolved step-scan FTIR spectroscopy enables the monitoring of conformational changes of proteins in the microsecond time scale (Bailey et al., 2002).

FTIR spectroscopy allowed to map out the nucleotide binding site of calcium ATPase (Liu and Barth, 2002). IR and FTIR spectroscopy are two of the only few methods suitable to monitor conformational changes of proteins under high pressure (Fig. 5.16; Dzwolak et al., 2002). FTIR spectroscopy on bacteriorhodopsin revealed a pre-melting conformational transition at 80 °C (Heyes et al., 2002). FTIR is also suitable to investigate the structure and hydration shell of protein molecules in organic solvents (Fig. 5.17; Costantino et al., 1995). Further, IR and FTIR spectroscopy was used for the characterization of irradiated starches (Kizil et al., 2002), and the determination of dihedral angles of tripeptides (Schweitzer-Stenner, 2002). Molecular changes of preclinical scrapie can be detected by IR spectroscopy (Kneipp et al., 2002). FTIR spectroscopy can serve as an optical nose for predicting odor sensation (van Kempen et al., 2002) and for chemical analysis of drinks (Coimbra et al., 2002; Duarte et al., 2002).

FTIR microscopy at a spatial resolution of 18 μm resolved single cells (Lasch et al., 2002). IR spectroscopy is also a tool for discrimination between different strains or types of cells (Gaigneaux et al., 2002).

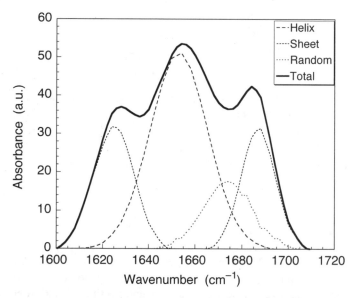

Fig. 5.18 Decomposition of a FTIR spectrum into three components corresponding to helical structure, sheets and non-regular structure, respectively. Percentages of structure content and structural changes, e.g., due to protein denaturation, are quantifiable

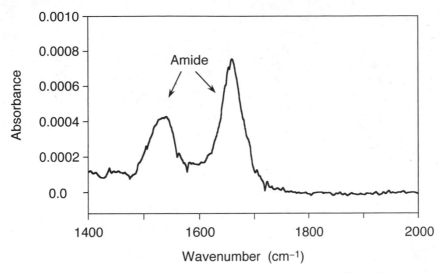

Fig. 5.19 FTIR spectrum of a single molecular monolayer of A126C sperm whale myoglobin (Jiang et al., 1996). The peaks around 1660 cm^{-1} and 1530 cm^{-1} correspond to the amide I and amide II bands, respectively. The spectrum was acquired with a BioRad FTIR spectrophotometer equipped with a TGS detector

Fig. 5.18 shows an example for the decomposition of a FTIR spectrum of a protein into the components corresponding to helical, sheet-like and random coil-like (non-regular) structures, respectively. Such decompositions can be calculated, e.g., by fitting a linear combination of the base spectra for the secondary structure components to the measured spectrum.

Fig. 5.19 illustrates the amazing sensitivity of FTIR spectroscopy. The sample was only two monolayers of a protein. Since at very low sample absorbances it is quite difficult to avoid the sharp lines of water-vapor absorption, these measurements were taken in a nitrogen-filled chamber at two different, very low concentrations of water, and later the water spectrum was subtracted. With this procedure, average artifact and noise levels were reduced to less than 0.00003 absorbance units.

6 Electron microscopy

6.1 Transmission electron microscope (TEM)

Transmission electron microscopy utilizes the wave properties of moving electrons to generate highly resolved images of specimens.

6.1.1 General design

In 1986 the Nobel prize in physics was awarded by one half to Ernst Ruska for his fundamental work in electron optics, and for the design of the first electron microscope (EM), and by one half to Gerd Binnig and Heinrich Rohrer for their design of the scanning tunneling microscope (see Chap. 7). In some aspects, the

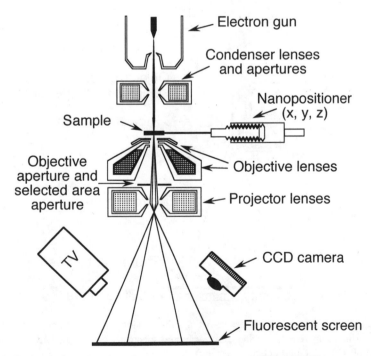

Fig. 6.1 Transmission electron microscope (see text on pp. 107 and 109)

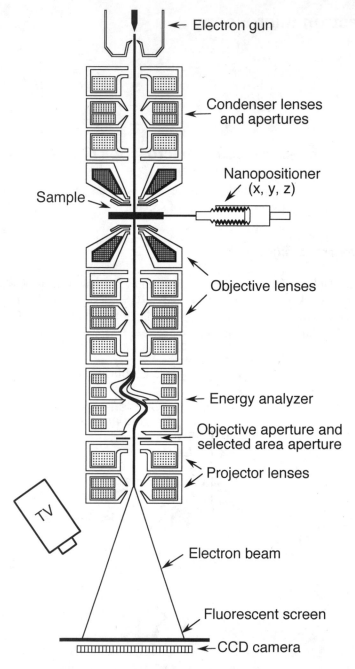

Fig. 6.2 A more complicated design of a transmission electron microscope with an analyzer which can remove inelastically scattered electrons (see, e.g., LEO Elektronen-mikroskopie GmbH, Oberkochen, Germany)

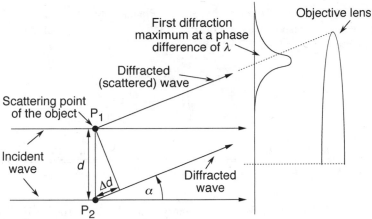

Fig. 6.3 In order to resolve the two points P_1 and P_2 of an object, the objective lens has to catch the first diffraction maximum of the two points. It appears in the direction where the diffracted (scattered) waves from the two diffracting (scattering) points have a phase difference, Δd, of one wavelength. Eq. (6.1) was derived from this condition. A typical objective lens has a bore of 2 mm and a focal length of about 1–2 mm

operation of a transmission electron microscope (TEM) is comparable with that of a slide projector (Figs. 6.1 and 6.2): Electrons from the electron gun pass through condenser lenses that focus the electrons onto the sample. The electron beam shines through the specimen. Objective lenses and projector lenses magnify the transmitted beam and project it onto the fluorescent viewing screen. Impact of electrons excites the screen and produces a visible magnified image of the sample. This image is recorded with various detectors, such as a CCD camera.

6.1.2 Resolution

Electron microscopes enable significantly greater magnification and greater depth of focus than conventional optical microscopes. High resolution TEMs permit spatial resolutions around 0.1 nm (1 Å) at acceleration voltages of 50–600 kV. Because of the wave nature of the electrons, the resolution limit, d, is given by the diffraction theory of coherent imaging:

$$d = \frac{\lambda}{n\sin(\alpha)} \, , \tag{6.1}$$

where λ, n, and α are the vacuum wavelength, index of refraction of the medium (=1 in TEMs), and aperture half angle of the objective lens, respectively (Fig. 6.3). The de Broglie relation provides the wavelength, λ, of the electrons:

$$\lambda = \frac{h}{mv} \, , \tag{6.2}$$

where $h = 6.6261 \times 10^{-34}$ J s, m, and v, are the Planck constant, electron mass, and electron velocity, respectively. For the relativistically high velocities of the electron beam we have to use Einstein's equations:

$$m = \frac{m_e}{\sqrt{1 - \dfrac{v^2}{c^2}}}, \quad E = mc^2, \tag{6.3}$$

and obtain:

$$\lambda = \frac{hc}{\sqrt{2E_0\Delta E + (\Delta E)^2}} = \frac{hce^{-1}}{\sqrt{2m_0c^2 e^{-1}V + V^2}} \approx \frac{1.24 \text{ nm kV}}{\sqrt{(1022 \text{ kV}) \cdot V + V^2}}, \tag{6.4}$$

where $e = 1.6022 \times 10^{-19}$ C is the elementary charge, $m_e = 9.1094 \times 10^{-31}$ kg the electron rest mass, $c = 2.99792 \times 10^8$ m s^{-1} the speed of light in vacuum, $E = E_0 + \Delta E$, $E_0 = m_e c^2$, $\Delta E = V \cdot e$ the kinetic energy of the electrons, and V is the applied acceleration voltage, typically 200 V – 200 kV. For a voltage of, e.g., 100 kV we find $\lambda = 0.0037$ nm. In contrast to most light microscopes, TEMs have small objective lens apertures of typically $\alpha = 1$–$2°$, and thus according to Eq. 6.1 the limit of spatial resolution is 0.1–0.2 nm in this example.

6.1.3 Electron sources

Thermionic electron guns (Fig. 6.4) and cold field emission guns (Fig. 6.5) are

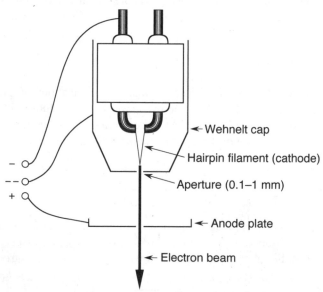

Fig. 6.4 Thermionic electron gun (e.g., Structure Probe, Inc., West Chester, PA)

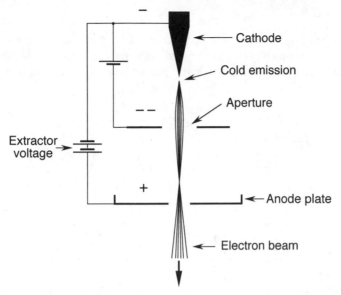

Fig. 6.5 Cold field emission gun focusable with an electrostatic lens comprised of two apertures with different electrostatic potentials

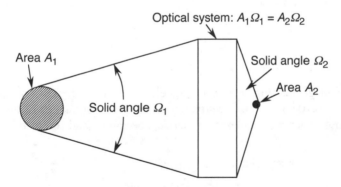

Fig. 6.6 In the production of an image of an object by an optical system, the product of area, A, and aperture solid angle, Ω, remains about constant

common electron sources. A considerable concern is the brightness and size of electron sources. Fig. 6.6 illustrates why this is important: For a light beam passing through an optical system, the product of area and aperture solid angle of radiation remains constant. Thus, a large source can be focussed on a small spot only by using a large aperture angle of the optical system. Considering the limited aperture angles of electron lenses, a source of small size and high brightness is required to obtain a sufficiently bright picture of the sample.

6.1.4 TEM grids

TEM grids (Fig. 6.7) should not get charged during measurement which would distort the electron path. Usually they are made from conductive chemically inert non-gassing materials suitable for high vacuum, such as platinum and platinum-iridium alloys.

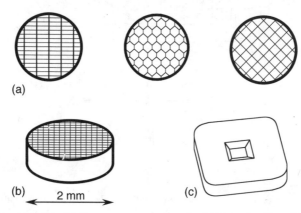

Fig. 6.7 TEM grids. (**a**), (**b**) Some common patterns of TEM grids made from a variety of materials, e.g., platinum, silver, tungsten, molybdenum, stainless steel, or titanium. Among these materials, platinum is the chemically most inert, but expensive. (**c**) Silicon nitride "grid" with a single window (from SPI Supplies, West Chester, PA)

6.1.5 Electron lenses

There are magnetic (Fig. 6.8 and 6.9), electrostatic (Fig. 6.5) and compound lenses (Figs. 6.10 and 6.11). Electron lenses have some similar characteristics like optical lenses, such as focal length, spherical aberration, and chromatic aberration.

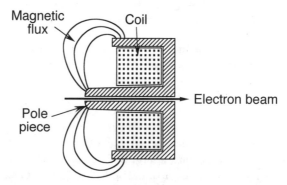

Fig. 6.8 Ernst Ruska's pohlschuh lens: the circular electromagnet is capable of projecting a precisely circular magnetic field in the region of the electron beam

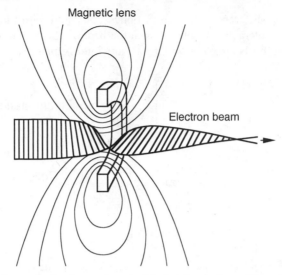

Fig. 6.9 Function of a magnetic electron lens (pohlschuh lens). One of Ernst Ruska's major achievements was the development of electron lenses

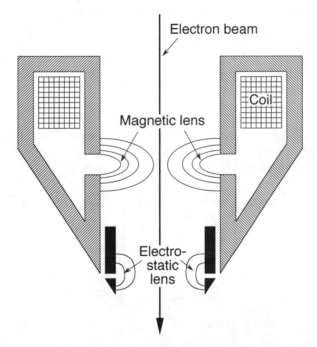

Fig. 6.10 Compound lens made from magnetic and electrostatic lenses: the magnetic field forces the electrons on spiral-shaped trajectories; the electric field further decreases the diameter of the electron beam. Additional coils may serve for the correction of spherical aberration (not shown)

One important difference of electron optics compared with photon optics is the mutual charge interaction of electrons in the beam. That is why electron optics is often designed for beam paths with few if any intermediate crossovers.

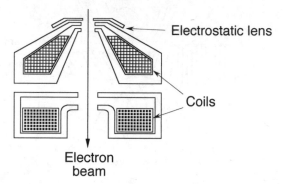

Fig. 6.11 Compound lens made from magnetic and electrostatic lenses

In the beam booster technique (LEO Elektronenmikroskopie GmbH, Oberkochen, Germany; Fig. 6.12) a high energy beam is generated, passed through the condenser column of the microscope, and then decelerated and passed through the sample. The high energy electrons are less affected by stray magnetic and electric fields. Also the propagation of the electrons in the column is independent from the selected electron probe energy.

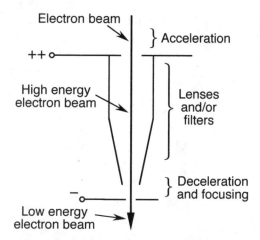

Fig. 6.12 Beam booster (LEO Elektronenmikroskopie GmbH, Oberkochen, Germany). The electrons are accelerated to a high energy, passed through condenser lenses and filters, and then decelerated prior to interacting with the sample. This technique largely protects the electron beam against stray magnetic fields in the column of the microscope

6.1.6 Electron-sample interactions and electron spectroscopy

There are different sources of chromatic aberration: (a) inelastic scattering of the electrons by the sample changes their energy (Fig. 6.13), and (b) the electrons leave the electron source with slightly different energies. The dispersion of electron energy is measured with energy filters (Fig. 6.14). Similar dispersive elements serve for the reduction of chromatic aberration, i.e. the selection of monochromatic electrons (Figs. 6.14 and 6.15)

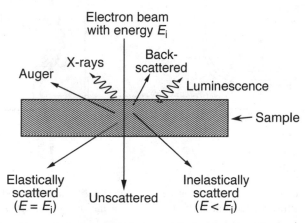

Fig. 6.13 Interaction of the electron beam with the sample. Inelastically scattered electrons have changed both direction and energy and may generate a diffuse contrast-reducing background image unless these electrons are eliminated by energy filtering (see Fig. 6.15). Elastically scattered electrons interfere with another and with unscattered electrons to produce a phase contrast image

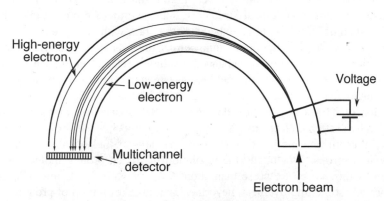

Fig. 6.14 Determination of electron energies. The voltage applied between the two hemispherical electrodes disperses the electrons with respect to their velocity. In order to record a full spectrum for a large range of electron energies, the applied voltage is swept

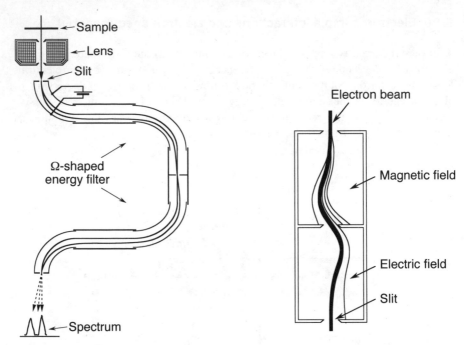

Fig. 6.15 Two types of dispersive elements for energy filtering of the electron beam to eliminate inelastically scattered electrons, removing the diffuse background and thereby enhancing contrast, or to perform a chemical analysis of the sample. *Left*: Ω-filter. *Right*: filter using magnetic and electric fields

Phase differences due to differences in the optical pathlength and electron scattering contribute to the contrast (Fig. 6.16). Often it is quite difficult to generate sufficient sample contrast at very high resolutions. A common method to visualize very small biological structures, such as single protein molecules, is negative staining: the sample is embedded in a stain with strong electron-optical properties (Fig. 6.17). Important innovations towards better contrast were the introduction of a technique for enhanced resolution (Haider et al., 1998) and the nanofabrication of solid-state Fresnel lenses for electron optics (Ito et al., 1998).

For biological samples a further important reason for the limitation of the resolution of TEMs is radiation damage, i.e. the destruction of the sample by inelastically scattered electrons. Since a certain number of electrons is necessary to obtain an image, this limit depends on the ratio of inelastically to elastically scattered electrons. Practically the resolution of frozen protein molecules is restricted by this reason to worse than about 5 Å. Negatively staining (Fig. 6.17) may provide some improvement, nevertheless atomic resolution of proteins is still beyond reach. It was suggested the theoretical possibility of a neutron microscope, for which the ratio of elastically to inelastically scattered particles may much better for isotope-exchanged proteins (Henderson, 1996). Another theoreti-

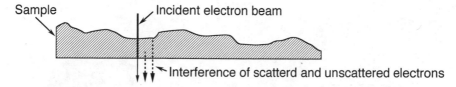

Fig. 6.16 Generation of amplitude contrast. The electron beam is weakened at different points to a different degree by scattering and interference: elastically scattered electrons, i.e., those which have changed direction but not energy, interfere with each other and with unscattered electrons to produce a phase contrast image

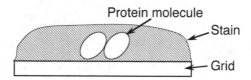

Fig. 6.17 Edge-on view of a negatively stained sample: the TEM senses volumes of lower density in the stain

cal possibility to overcome the problem of the decay of the structure of sample during the measurement might be the use of ultrashort electron flashes on deeply frozen samples: if the duration of the flash is shorter than the time of mechanical movement of the protein molecule, its chemical decomposition would affect the obtained micrograph to a lesser degree.

6.1.7 Examples of biophysical applications

Fig. 6.18 demonstrates the resolution power of TEM for large protein complexes (Roseman et al., 1996; White et al., 1997; Ranson et al., 1998; Rye et al., 1999; Saibil, 2000a). Clearly differences between two conformations of GroEL/GroES are resolved. The TEM structure is consistent with the crystal structure.

Electron microscopy resolved the structure of the bacteriophage Φ29 packaging motor (Simpson et al., 2000) and visualized the filamentous phage pIV multimer (Linderoth et al., 1997). Electron microscopy contributed to the understanding of conformational changes connected with the opening of an ion channel through a membrane (Saibil, 2000b), and with connexin trafficking (Gaietta et al., 2002).

In groundbreaking experiments Terry G. Frey and coworkers succeeded in the 3D-visualization of cell organelles using electron tomography. In this method the three-dimensional structure is calculated from a series of electron micrographs of samples tilted over a range of angles (Dierksen et al., 1992; Perkins et al., 1997a, 1997b; Frey and Mannella, 2000).

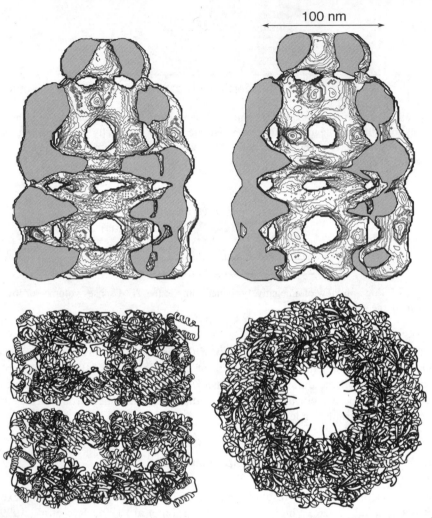

Fig. 6.18 *Top:* electron micrograph of two conformations of GroEL/GroES at 30 Å resolution (Roseman et al., 1996; White et al., 1997; Ranson et al., 1998; Rye et al., 1999; Saibil, 2000a). *Bottom:* the X-ray crystal structure of GroEL at 2.4 Å resolution for comparison (Braig et al., 1994; Boisvert et al., 1996). The latter figure part was generated using MOLSCRIPT (Kraulis, 1991)

6.2 Scanning transmission electron microscope (STEM)

In contrast to TEMs, scanning transmission electron microscopes use an electron beam with only a few Å or nm diameter to scan the sample area (Fig. 6.19). The resolution is generally limited by the diameter of the electron beam at the location of the sample and radiation damage.

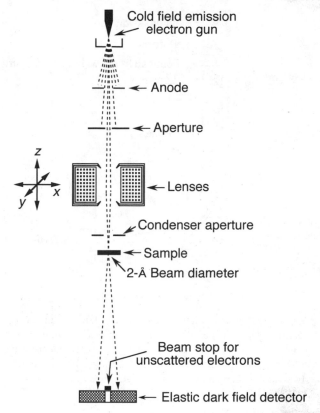

Cold field emission
electron gun

Anode

Aperture

Lenses

Condenser aperture

Sample
2-Å Beam diameter

Beam stop for
unscattered electrons

Elastic dark field detector

Fig. 6.19 Scheme of a scanning transmission electron microscope: The objective lens focuses the electron beam onto a small sample section. Scattered electrons are detected with the elastic dark field detector. The STEM image is generated by moving the focussed beam over the specimen

Although the STEM was pioneered already in the thirties of last century, mainly by adding scan coils to a TEM (von Ardenne, 1940), significant developments have taken place in the last years: electron optics has been significantly improved and the resolution increased by several orders of magnitude. Nowadays high resolution STEMs offer unprecedented capability for the characterization of biomolecules, allowing structure to be determined with up to sub-nm resolution.

Similarly to TEMs, the STEM can employ various energy filtering techniques for chemical analysis and improvement of resolution, e.g., by removing unscattered electrons in inelastic dark field imaging (Fig. 6.20). Many STEM have both capabilities, elastic dark field imaging (Fig. 6.19) and inelastic dark field imaging (Fig. 6.20). A third mode is bright field detection where electrons are collected through a small aperture placed on the optical axis and an energy

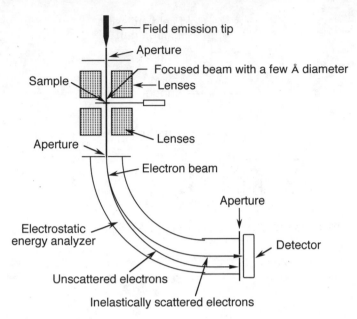

Fig. 6.20 Example of a scanning transmission electron microscope with an energy filter: inelastically scattered electrons, i.e., those which have changed both energy and direction upon interaction with the specimen, are collected yielding the inelastic dark field image. Electrons with different energies are separable by their trajectories with different curvatures in the electric field applied perpendicular to the flight direction

filter removes those electrons that have lost energy, i.e. low-angle elastically scattered and unscattered electrons are collected to produce the image.

7 Scanning probe microscopy

Scanning probe microscopes generate a highly-resolved image of the specimen by scanning it with a small mechanical, electrical, optical, thermal, or other probe.

7.1 Atomic force microscope (AFM)

The AFM was invented by Gerd Binnig, Christoph Gerber, and Calvin F. Quate in the mid-eighties (Binnig et al., 1986), and is one type of the so-called scanning

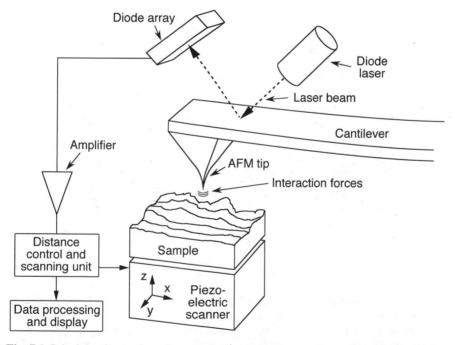

Fig. 7.1 Principle of operation of an atomic force microscope. A very sharp tip attached to a tiny cantilever probes the sample surface. An optical system comprised of diode laser and detector, e.g., a diode array or a position-sensitive diode, senses the bending of the cantilever and thereby the distance-dependent tip-sample interaction force. For scanning the surface, the sample is moved by the piezoelectric scanner (Binnig et al., 1986)

probe microscopes (SPMs) which also include scanning tunneling microscope (STM; Sect. 7.2; Binnig et al., 1982a, 1982b, 1983; Binnig and Rohrer, 1987), scanning near-field optical microscope (SNOM; Sect. 7.3), scanning thermal microscope (SThM; Sect. 7.4), and the scanning ion conductance microscope (SICM; Sect. 7.4). The AFM is used in both industrial and fundamental research to obtain atomic-scale images of metal surfaces and nanometer-scale images of the three-dimensional profile of the surface of biological specimens. It is a very useful tool for determining the size and conformation of single molecules and aggregates adsorbed on solid surfaces. The AFM scans the sample with a tiny tip mounted on a small cantilever (Fig. 7.1). It measures the small force of interaction between tip and sample surface by sensing the reflection changes of a laser upon cantilever movement caused by interaction with the sample. An image of the sample surface relief is recorded using piezoelectric translation stages that move the sample beneath the tip, or the tip over the sample surface, and are accurate to a few Å.

Note the similarity of the AFM (Fig. 7.1) to the stylus profilometer (Fig. 7.2) and to the STM (Fig. 7.19). Actually, the idea of AFM is based on the design of stylus profilometers, but the AFM can reveal the sample relief with subnanometer resolution.

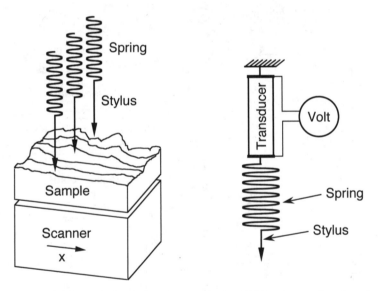

Fig. 7.2 Stylus profilometer for comparison with the AFM. A set of styli probes the sample which is drawn below the set of styli. The small motions of the styli are transformed into an electrical signal by linear, variable transducers. Step heights of down to a few 10 nm are resolvable

The force of tip-sample interaction (Fig. 7.3) has a magnitude of typically only a few pN – nN. That is why the cantilever must have a small mass, and the weight-bearing parts of the AFM (Figs. 7.4 and Fig. 7.5) have to be rigid and equipped with a good vibrational damping. Three support posts in the design of Fig. 7.4 reduce wobbling. The xyz-translation stages for coarse adjustment of the piezoelectric scanner with the sample on top are engineered for little wobbling as well. The whole AFM is placed on a rubber support preventing transmission of high frequency vibrations from the laboratory (not shown). A low force of interaction is crucial for high resolution force microscopy on soft biological specimens. Low spring constants of the cantilever may facilitate this purpose at the expense of resolution, but the most common way of gentle measurement is to reduce the intensity and duration of contact by oscillating the cantilever, as will be explained later (Fig. 7.13).

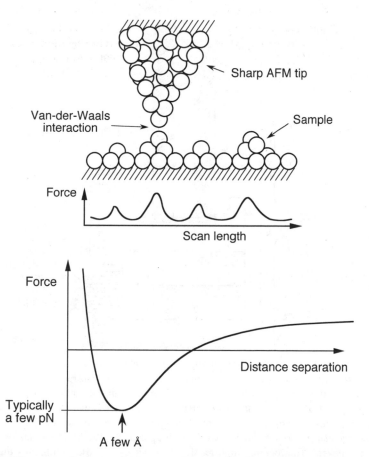

Fig. 7.3 When a small tip approaches a surface, it experiences the van-der-Waals force which is attractive at a distance of a few Å, but repulsive at very short distances (see, e.g., Chap. 3 in Nölting, 1999b). Additional Coulomb forces may play a role when the AFM tip was charged

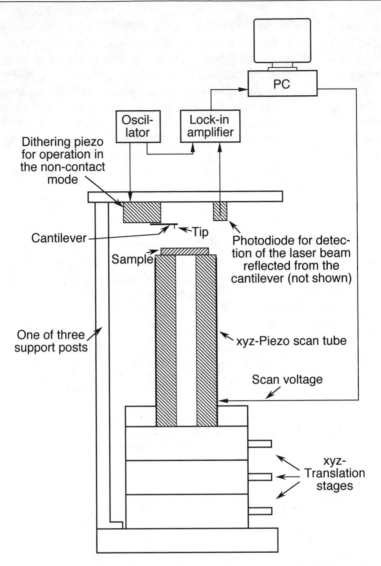

Fig. 7.4 Design of an AFM. The sample spot of interest is positioned near the tip by coarse xyz-translation stages. The piezoelectric scanner (see also Fig. 7.5) then heightens the sample position further till the tip starts to interact with the sample. It allows motion control of the tip with subnanometer precision. A photodiode detects the reflection changes of a laser beam from the cantilever upon approach of tip to sample. In this example, the cantilever is mounted to a dithering piezo element which excites oscillations of the cantilever. The lock-in amplifier detects changes of these oscillations due to tip-sample interactions. The sample surface is scanned by sample movement in horizontal direction by the piezoelectric scanner. The scanner also adjusts the relative height of the cantilever during scanning to avoid crashes of the tip with the sample surface. Such crashes can damage the tip and then cause artifacts (see Fig. 7.6)

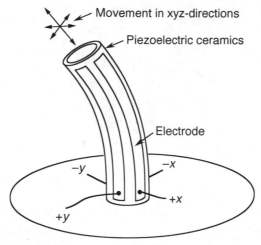

Fig. 7.5 Principle of operation of the xyz-piezoelectric scanner, a ceramic positioning device which changes its size in response to a change in applied voltage. A voltage change at the x- or y- electrodes causes bending in the horizontal plane; contraction and expansion are generated by simultaneous application of x- and y-voltage

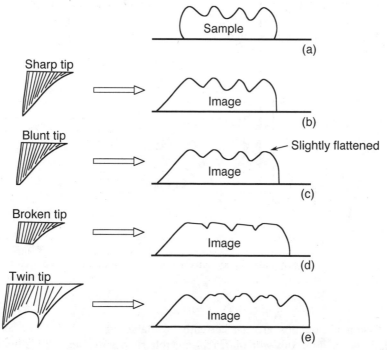

Fig. 7.6 Artifacts caused by different shapes of the AFM tip. Blunt tips and broken tips give rise to an seemingly flattened sample relief which may be difficult to recognize as an artifact

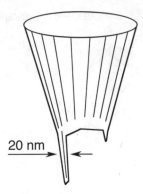

Fig. 7.7 Tungsten tip, made by sharpening a tungsten wire by dragging it over a plate coated with alumina. This self-made tip may also be used for STMs (see Sect. 7.2)

Obviously a robust and sharp single tip is essential for this method. Typical apex radii of commercial tips are $\approx 10\text{--}20$ nm. Fig. 7.6 depicts common types of artifacts observed when using worn out tips, broken tips, or probes with more than one tip.

One can make tips themselve by grinding a tungsten wire on a sheet covered with alumina (Fig. 7.7). These tips are also suitable for STM (Sect. 7.2), but the tip shape is not very reproducible and tungsten is not very hard.

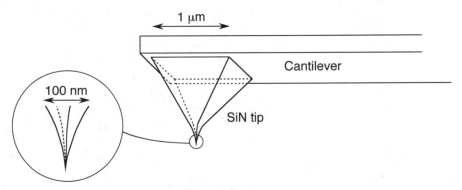

Fig. 7.8 Sharpened pyramidal silicon nitride tip. SiN is extremely hard, and tips can be engineered with radii of only a few nm

Silicon nitride tips are the currently available tips with highest robustness (Fig. 7.8). Sharper tips with quite reproducible shape are made from silicon which is relatively fragile, however (Fig. 7.9a). Even sharper tips for application on samples with particularly deep structures are manufactured by attaching a high density carbon fiber to a silicon tip (Fig. 7.9b). Sharp cantilevers for the examination of very rough surfaces (Fig. 7.10) and cantilevers with trigonal design for the purpose of high resistance against torsion (Fig. 7.11) are supplied, e.g., by Olympus Optical Co. (Tokyo).

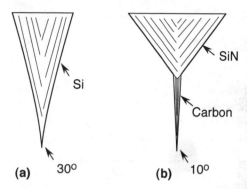

Fig. 7.9 (**a**) AFM tip made from silicon. (**b**) Silicon tip with a high density carbon fiber attached to it

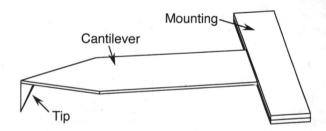

Fig. 7.10 Sharp cantilever geometry for very rough samples (e.g., Olympus Optical Co., Tokyo)

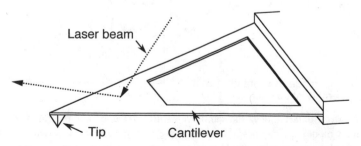

Fig. 7.11 A trigonal design of the cantilever (e.g., Olympus Optical Co., Tokyo) causes a better stability against torsion, compared with rod-shaped cantilevers

The life of the AFM tip decreases very rapidly with applied force (Fig. 7.12). High aspect-ratio tips made from silicon or carbon fiber are generally less durable than low aspect-ratio tips made from silicon nitride.

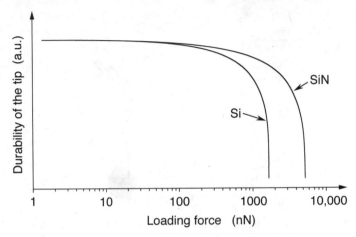

Fig. 7.12 Example for the wear of two AFM tips due to surface load

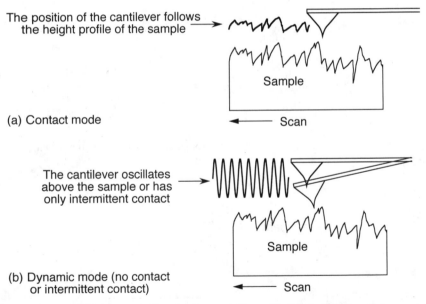

Fig. 7.13 **(a)** Contact mode: the cantilever follows the height profile of the sample. **(b)** Dynamic mode: the cantilever has only intermittent contact or oscillates above the sample. In the latter mode, oscillations are excited by a piezo crystal and the forces between tip and sample are very small. This mode permits truly atomic resolution (Giessibl, 2000)

There are two common modes of operation of AFMs (Fig. 7.13): the contact mode and the dynamic force mode. In the contact mode, the probe tip is in continuous contact with the sample surface. The force the cantilever exerts on the substrate in contact mode may perturb the surface of soft biological materials. In the gentler dynamic mode, the probe tip only oscillates up and down as it is scanned over the sample surface. Two sub-modes may be distinguished for the dynamic force mode, the non-contact sub-mode in which the distance between tip apex and sample surface is always larger than the van-der-Waals distance, and the tapping sub-mode in which the tip has intermittent contact.

As pointed out, a high degree of protection against external high-frequency vibrations is obviously crucial for the operation scanning probe microscopes with atomic resolution. Fig. 7.14 shows a further solution to this problem. Here the AFM is made from very thick and short plates of steel and the AFM is placed on three rubber balls that do not transmit fast vibrations. Another technique of efficient vibrational damping is to hang the AFM on a rubber string (Fig. 7.15).

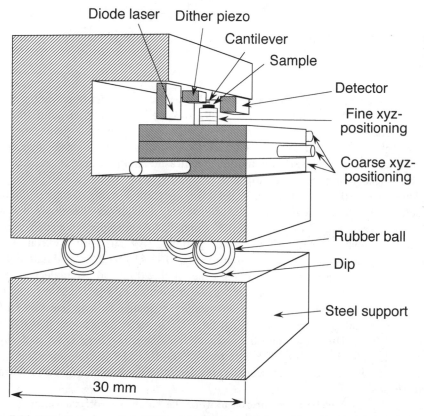

Fig. 7.14 Robust design of an AFM with atomic resolution. The vibrational damping is attained by a very rigid construction and an elastic support in form of three rubber balls

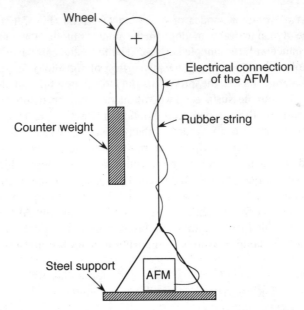

Fig. 7.15 "Hanging AFM": vibrational isolation of an AFM by hanging it on a rubber string

Important biological applications of AFMs were the direct observation of the structure of DNA (Lindsay et al., 1989) and the monitoring of actin filament dynamics in living cells (Henderson et al., 1992). The direct visualization of a DNA glycosylase searching for damage shows that the glycosylase interrogates DNA at undamaged sites by introducing drastic kinks (Chen et al., 2002b). Intramolecular triplex DNA formation results in a kink in the double helix path (Tiner et al., 2001). A sharp DNA bend is induced by binding of integration host factor (IHF) to the region between the upstream regulatory sequence and the promoter sequence (Seong et al., 2002). Single DNA molecule force spectroscopy can discriminate between different interaction modes of small drug molecules with DNA by measuring the mechanical properties of DNA and their modulation upon binding of small drug molecules (Krautbauer et al., 2002) and dye molecules (Kaji et al., 2001). A decrease of the ionic strength from 50 mM to 1 mM resulted in a change of the number of nodes (crossings of double helical segments) of a supercoiled 3000-bp piece of DNA from a 15 to one or two nodes (Cherny and Jovin, 2001). High resolution fluorescence imaging of λ-phage DNA molecules, intercalated with the dye YOYO-1, by a SNOM/AFM (SNOM, scanning near-field optical microscope; see Sect. 7.3) resolved the distribution of the dye (Kim et al., 2001).

AFM proved to be a very useful tool for the study of proteins, yielding some unique insights into structure and physical properties: β-Lactoglobulin forms fine-stranded aggregates at pH 2 with the diameter of strands being ca. 4 nm (Ikeda and

Morris, 2002). AFM technology was used to map out the electrostatic potential of the transmembrane channel OmpF porin (Fig. 7.16; Philippsen et al., 2002). AFMs gave crucial topological information of blood cell adhesion on different sensor materials (Hildebrand et al., 2001). Ac-GWWL(AL)nWWA-Etn peptides induce the formation of extremely ordered domains in some biologically relevant membranes (Rinia et al., 2002). The heads of bacteriophage ΦKZ and T4 have different compressibilities (Matsko et al., 2001). Atomic force microscopy resolved fusion pores in the apical plasma membrane in live pancreatic cells (Cho et al., 2002) and visualized the growth of Alzheimer's β-amyloid-like fibrils (Goldsbury et al., 2001). Cardiac muscle and skeletal muscle exhibit different viscous and elastic properties as determined by atomic force microscopy. Cardiac cells are stiffer (elastic modulus = 100 ± 11 kPa) than skeletal muscle cells (elastic modulus = 25 ± 4 kPa; see Mathur et al., 2001). Atomic force microscopy allowed to visualize the structure of biomolecules, e.g., the native chaperone complex from *Sulfolobus solfataricus*, in solution under physiological conditions providing a nanometer resolution topographic image of the sample (Valle et al., 2001). It is also an excellent technique to study the initial events of mutual cell adhesion (Razatos, 2001). An AFM image of a monomolecular film of bovine serum albumin shows individual monomers and dimers (Fig. 7.17; Gunning et al., 1996; Morris et al., 1999).

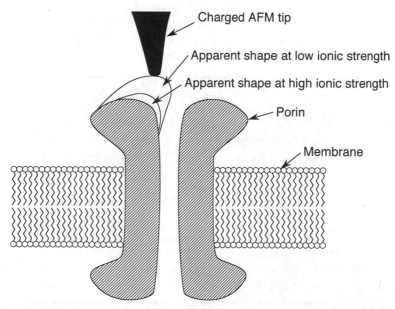

Fig. 7.16 Imaging the electrostatic potential of the transmembrane channel OmpF porin (Philippsen et al., 2002). Different apparent shapes of the porin are observed at different ionic strengths. These differences reflect changes of the electrostatic potential which is experienced by the charged tip of the AFM

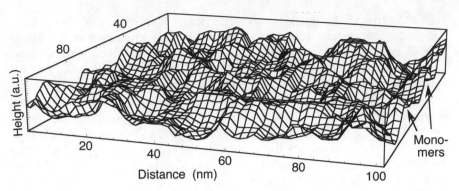

Fig. 7.17 AFM image of a monomolecular film of the protein bovine serum albumin (BSA, $M_{w,monomer}$ = 66 kDa) adsorbed at an oil/water interface (Gunning et al., 1996; Morris et al., 1999). Individual monomers and dimers of BSA can be seen

AFMs are also very useful for the manipulation of macromolecules: proteins may physisorb to the AFM tip and can then be lifted and manipulated (Fig. 7.18). The sensitivity of the AFM cantilever, to forces in the pN range, was exploited to measure folding-unfolding forces within single protein molecules and breakaway forces between different biomolecules (Jiao et al., 2001; Allison et al., 2002). Atomic force microscopy has yielded tantalizing insights into the dynamics of protein self-assembly and the mechanisms of protein unfolding (Furuike et al., 2001; Yip, 2001). For further, similar applications of AFM technology see Chap. 8.

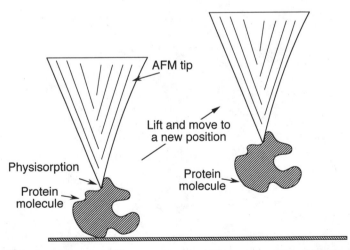

Fig. 7.18 Manipulation of a protein molecule with an AFM: The tip is lowered till it touches the macromolecule. Due to the attractive action of the van-der-Waals interaction, the macromolecule sticks to the tip and can be lifted and moved to a different place

7.2 Scanning tunneling microscope (STM)

In 1986 Gerd Binnig and Heinrich Rohrer were awarded the Nobel Prize for Physics for the groundbreaking invention of the STM. It was the first member of the family of scanning probe microscopes (SPM) that can characterize surface morphology with atomic resolution. In contrast to the AFM, its principle of operation (Fig. 7.19) requires electrically conductive samples. A sharp tip located on a flexible cantilever is used to probe the distance between the tip and sample surface, as judged by the tunneling current (Fig. 7.20). Since the tunneling current also depends on the chemical nature of sample and tip, the STM also serves for characterization of electronic properties of solid samples.

Significant complications on the way towards higher resolution of proteins are the undesired excitation of the soft biological material by the high current of STMs, typically pA−nA, and the distribution of conductivity within the sample distorting the image (Fig. 7.21). Low currents and stable attachment of the sample to the support are required for high resolution images of biological macro-molecules.

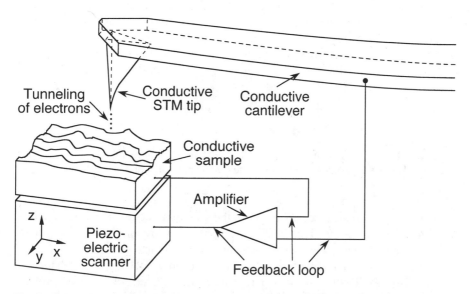

Fig. 7.19 Principle of operation of STMs. A finely sharpened electrically conductive tip is first positioned within about 1 nm of the sample by mechanical translation stages (not shown) and the piezoelectric scanner. At this small separation, electrons tunnel through the gap between tip and sample (Fig. 7.20). The tunneling current depends on the applied bias voltage between tip and sample, the distance, the tip shape, and the chemical compositions of sample and tip. The feedback loop ensures constant height or constant current. Tunneling current and feedback voltage are a measure of surface morphology and composition (Binnig et al., 1982a, 1982b, 1983; Binnig and Rohrer, 1987)

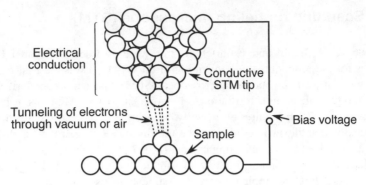

Fig. 7.20 When the distance between conductive tip and conductive sample is lowered to a few Å, electrons can traverse the gap with some transmission probability. The STM measures not purely distance like the AFM, but the local density of electronic states

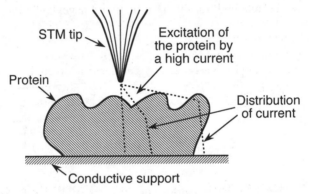

Fig. 7.21 Different possible current paths in STM measurements of extended biological structures lower the resolution and complicate the interpretation of data

Similar as in atomic force microscopy, the quality of the tip is crucial for a high resolution (Figs. 7.7–7.11, 7.22). Commonly tips are micromachined and/or electrochemically etched and have apex radii of 5–100 nm.

STM images of a 1:1 mixture of K344C cytochrome P450$_{cam}$ / putidaredoxin adsorbed on gold (111) showed a regular array of pairs of the two proteins (Djuricic et al., 2002). I21C/E25C plastocyanin essentially maintains its native redox properties upon immobilization onto a gold substrate as shown by the full potentiostatic control of the electron transfer reaction during STM imaging (Andolfi et al., 2002). Scanning tunneling microscopy demonstrated that the otherwise nearly linear mammalian metallothionein-2 molecule bends by about 20 degrees at its central hinge region between the domains in the presence of ATP (Maret et al., 2002). Electrochemical scanning tunneling microscopy on thiol-derivatized DNA immobilized on a gold (111) single crystal surface showed

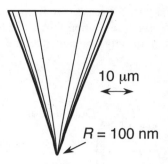

Fig. 7.22 An etched tungsten tip for STM (see also Figs. 7.7–7.11)

potential-dependent orientation changes of the DNA in the potential range from 200 to 600 mV (Zhang et al., 2002b). A STM study of morphology and electron transport features in cytochrome c offers evidence for sequential discrete electron-tunneling effects (Khomutov et al., 2002). Tunneling in proteins adsorbed onto a conductive substrate may depend on the applied potential (Facci et al., 2001). The resistance of a single octanedithiol molecule is 900 MΩ (Cui et al., 2001). The ability to site-specifically introduce cysteine residues and to engineer tags, such as histidine tags and biotin-acceptor peptides, allow the creation of ordered immobilized protein structures that can be characterized both electrochemically and topographically by using scanning probe microscopy and cyclic voltammetry (Gilardi et al., 2001).

7.3 Scanning nearfield optical microscope (SNOM)

7.3.1 Overcoming the classical limits of optics

SNOMs (Figs. 7.23–7.25), also known as NSOMs, utilize a light source with a diameter smaller than the wavelength of the light (Synge, 1928; Ash and Nicholls, 1972; Pohl et al., 1984; Betzig et al., 1986, 1991, 1992; Toledo-Crow et al., 1992; Williamson et al., 1998; Egawa et al., 1999; Heimel et al., 2001). By means of this technological innovation they achieve a resolution which may be well beyond the resolution limit, d, of classical Abbe-Fourier optics (see also Sect. 6.1.2):

$$d = \frac{\lambda}{n \sin(\alpha)}, \tag{7.1}$$

where λ, n, and α are the vacuum wavelength, refractive index of the medium between sample and objective lens, and half angle of aperture, respectively. For visible light with $\lambda = 500$ nm, $n = 1.6$, and α near 90°, we obtain a resolution limit of classical optics of about 300 nm. Using UV light and image processing can yield improvement beyond this, but it is clear from Eq. 7.1, that classical optics can hardly penetrate the 100-nm resolution barrier. SNOMs have been the first optical microscopes that significantly overcame the limit of Eq. 7.1.

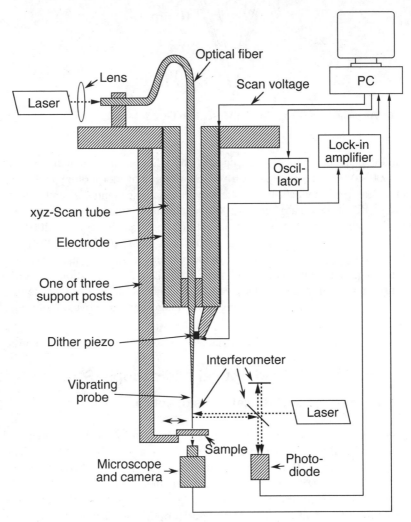

Fig. 7.23 A design of a SNOM. An optical probe emits light from points above the 2D sample. The subwavelength probe tip takes advantage of the physical effect of optical tunneling. In this SNOM it is made using a metal coated tapered glass fiber: a first taper for the probe was manufactured by melt-drawing, and a second taper at the very end of the probe was etched. The damping of the horizontally vibrating SNOM probe caused by shear forces is taken as a measure for the distance from the sample. A shear force feedback loop which involves an interferometric measurement of the horizontal position of the probe tip keeps it at constant height (Betzig et al., 1992). Near-field optical properties of the sample surface are mapped out by scanning each point within a certain area

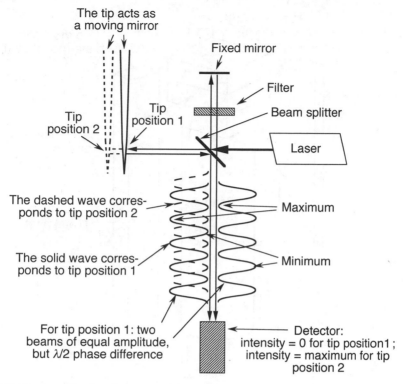

Fig. 7.24 Design of an interferometer used for the detection of shear force of the SNOM tip. The principle of operation is based on the extinction of light at the position of the detector due to the interference of the two incident light beams of equal amplitude when the probe tip (movable mirror) is in position 1. A small perturbation of the position of the tip causes a small deviation from $\lambda/2$ of the phase shift of the two beams inciding on the detector which leads to a non-negligible detector signal. The light filter in the fixed path ensures an equal amplitudes of the two interfering beams

Generally the working principle of SNOMs is as follows (Figs. 7.23–7.25): The subwavelength light source is positioned a few μm above the specimen surface with the help of mechanical coarse translation stages and a piezoelectric fine translation device, e.g., a piezoelectric scan tube. Transmission is measured below the specimen. An image of the specimen surface is obtained by moving the light source (or the sample in some designs) with the help of the piezoelectric fine translation device in horizontal direction. To avoid crashes with the sample, in many SNOMs the light source is oscillated over the specimen surface and damping of oscillations due to source-sample interactions detected. The piezoelectric fine translation device somewhat lifts the subwavelength light source (or lowers the sample) when damping increases. Near-field optical excitation of the sample can be seen as a dipole-dipole energy transfer (Sekatskii and Dietler, 1999).

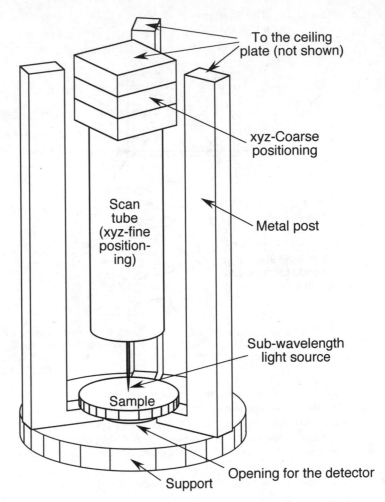

Fig. 7.25 A design of a SNOM with a xyz-coarse positioning and a xyz-fine positioning

There is virtually no resolution limit of SNOMs as long as one can manufacture light sources of sufficient small size (see next section) and detect very small intensity differences of light passing through thin sample layers.

7.3.2 Design of the subwavelength aperture

The most common methods to manufacture subwavelength apertures are (a) adiabatic pulling of an optical fiber during heating (Betzig et al., 1991; Williamson and Miles, 1996; Figs. 7.26a and 7.27), (b) etching (Muramatsu et al., 1999), in particular tube etching (Turner, 1983; Stöckle et al., 1999a; Figs. 7.26b and 7.28), and (c) microfabrication (Schurmann et al., 2000; Mitsuoka et al., 2001), e.g., by ion beam milling (e.g., Veerman et al., 1998).

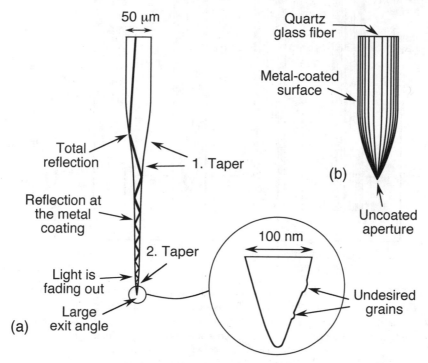

Fig. 7.26 (**a**) Melt-drawn probe: in the upper part of the probe, the light is reflected from the walls by total internal reflection (TIR). TIR is the phenomenon involving the reflection of all the incident light off a boundary when both (i) the light is traveling in the more optically dense medium and approaching the less optically dense medium and (ii) the angle of incidence is greater than the so-called critical angle. (**b**) Etched probe. A desirable high brightness is obtained by a large cone angle. An optical aperture is formed by rotational evaporation of the etched fiber with an opaque metal, e.g., aluminum

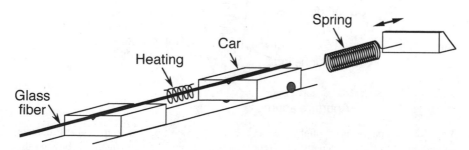

Fig. 7.27 Fabrication of a probe by adiabatic pulling of an optical fiber during heating (melt-drawing; see, e.g., Williamson and Miles, 1996)

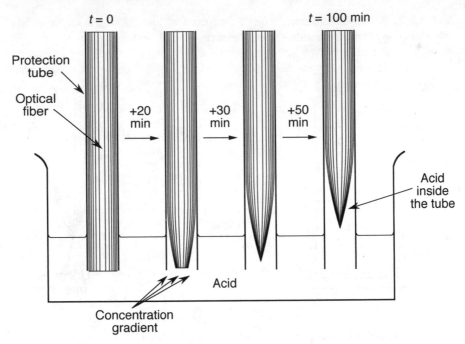

Fig. 7.28 Tube-etching of a quartz glass fiber with hydrofluoric acid (Turner, 1983; Stöckle et al., 1999a, 1999b): the tip forms due to concentration gradients of acid and desolved fiber. The tube serves also for suppression of convection of the acid. These tube-etched tips are inexpensive and have large cone angles permitting high light throughputs

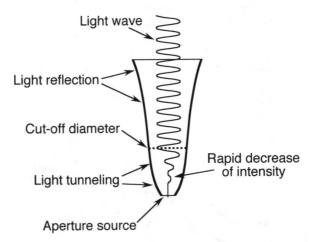

Fig. 7.29 Propagation of a light wave in a SNOM probe tip. Up to the cut-off diameter (about equal to the wavelength), the light travels with only little attenuation. Beyond this critical diameter, the light intensity very rapidly falls off

Usually the aforementioned probe tips are coated with a metal. Since the light transmission of the tip dramatically depends on the distance from the cut-off region to the aperture (Fig. 7.29), the optical throughput of etched tips with cone angles around 30° is, in general, 2–3 orders of magnitude better than that of heat-pulled fiber probes. Even brighter sources can be microfabricated (Fig. 7.30).

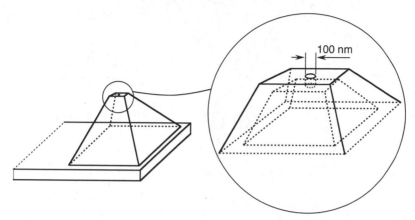

Fig. 7.30 SNOM tip made from silicon nitride. It was manufactured by using photo-lithography, potassium hydroxide etching, and electron beam nanolithography (Zhou et al., 1998, 1999)

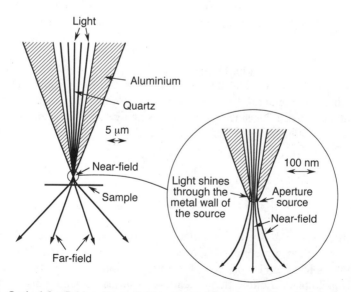

Fig. 7.31 Optical far-field and near-field in the vicinity of a small optical tip for a SNOM. Since the light can transmit through thin layers of metal, the diameter of the light beam in this design cannot be less than a few 10 nm

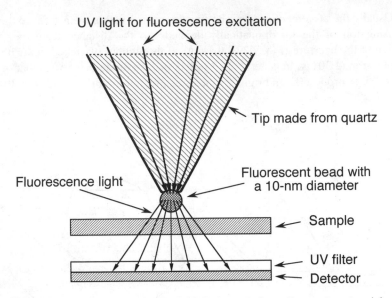

Fig. 7.32 A 10-nm sized light source for SNOMs made from fluorescent material. UV light excites fluorescence of the 10-nm sized bead at the end of the probe. The transmission of the fluorescence light through the sample is sensed by a detector which is covered with an UV-absorbent layer. Antioxidants may be added to the fluorescent material to enhance stability and life. The resolution for thin-layered samples enabled by this type of light sources depends on the size of the fluorescent bead and the detectability of small absorbance differences of the fluorescent light

In the aforementioned methods of source fabrication, a limitation for the source size is the transmission of light through thin layers of metal (Fig. 7.31). Sources of 10 nm diameter could hardly be made just by machining an opening in a metal plate or in an optical fiber since light would significantly shine through the walls of the opening. Smaller light sources involve the excitation of small fluorescent particles (Fig. 7.32).

7.3.3 Examples of SNOM applications

An important applicability is seen in cell biology, microbiology (Meixner and Kneppe, 1998), and proteomics (Gao et al., 2001). SNOM imaging visualized domains of photosystem II core complex bound to lipid monolayers (Trudel et al., 2001). Topographic, friction, fluorescence, and surface potential distributions for a Langmuier-Blodgett film can simultaneously be observed using a SNOM-AFM with a thin step-etched optical fiber probe (Horiuchi et al., 1999). Fig. 7.33 demonstrates a 50-nm resolution.

200 nm

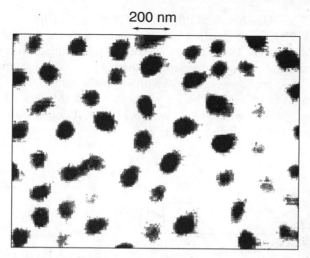

Fig. 7.33 Absorption of latex beads with approximately 100 nm diameter taken with a SNOM at a resolution of about 50 nm (OMICRON, Taunusstein, Germany)

7.4 Scanning ion conductance microscope, scanning thermal microscope and further scanning probe microscopes

Important scanning probe microscopes are also the magnetic force microscope, scanning Hall probe microscope (Chang et al., 1992a, 1992b), friction force microscope (Fig. 7.34; Howald et al., 1995), scanning ion conductance micro-

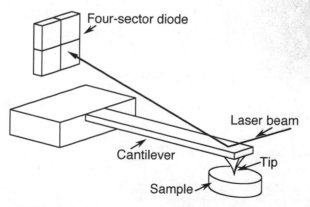

Fig. 7.34 Example of a friction force microscope: the detector has sectors in both vertical and horizontal direction so that the torsion of the cantilever can be estimated and a frictional (lateral) force be calculated (see, e.g., Howald et al., 1995)

scope (Fig. 7.35; Hansma et al., 1989; Korchev et al., 1997, 2000a, 2000b; Stachelberger, 2001; Bruckbauer et al., 2002a), and scanning thermal microscope (Fig. 7.36; Mills et al., 1998, 1999). There are two common types of thermal probes for scanning thermal microscopes: thermocouples and thermal resistors. The thermocouple probe in Fig. 7.36 involves two dissimilar metal wires which bisect over the top of a blunt silicon nitride pyramid (Mills et al., 1998). A potential biophysical application is the elucidation of local heating effects in biological tissue due to cell metabolism.

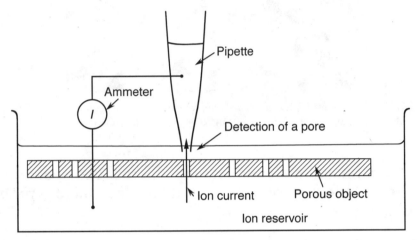

Fig. 7.35 Detection of membrane pores by a scanning ion conductance microscope. The probe is a nanopipette filled with electrolyte solution. The current between pipette and ion reservoir starts increasing when the pipette tip approaches a pore. The technique permits 100-nm resolution characterization of distribution and sizes of pores

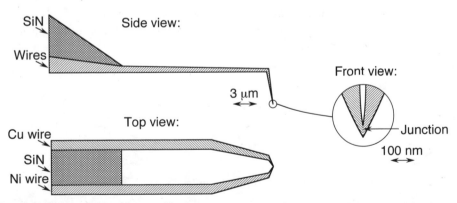

Fig. 7.36 Design of the tip of a scanning thermal microscope (Mills et al., 1998, 1999). A submicrometer-sized Cu/Ni thermocouple at the end of the cantilever detects the thermal microenvironment

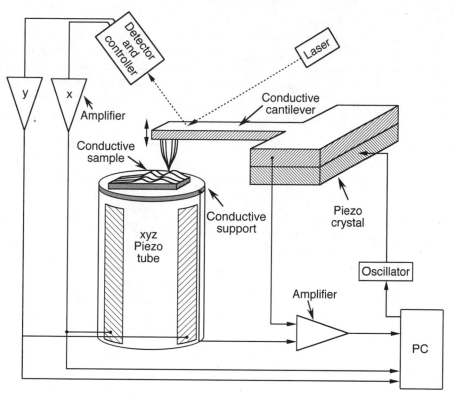

Fig. 7.37 STM with simultaneous AFM capability: AFM and STM sense different physical properties. The combined information provides greater insight into the chemical nature of the sample ant its physico-chemical properties

The STM with simultaneous AFM capability (Fig. 7.37) provides simultaneously information about surface relief and chemical composition of the specimen.

8 Biophysical nanotechnology

8.1 Force measurements in single protein molecules

Atomic force microscope (AFM)-related techniques can induce and monitor the unfolding of single protein molecules. Experiments on the protein titin, which is a main component of skeletal muscles (Figs. 8.1–8.3), revealed that the force for unfolding of its individual domains with cross sections of less than 5 nm^2 is of the order of 100–300 pN and dependent on the pulling speed (Rief et al., 1997; Gaub and Fernandez, 1998; Carrion-Vasquez et al., 1999). A similar investigation on bacteriorhodopsin showed that its helices are anchored to the bacterial membrane with 100–200 pN (Fig. 8.4; Oesterhelt et al., 2000). Similarly, single-molecule force spectroscopy on spider dragline silk protein molecules revealed that the molecule unfolds though a number of rupture events, indicating a modular structure within single silk protein molecules (Oroudjev et al., 2002). The minimal unfolding module size of 14 nm indicates that the modules are composed of 38 amino acid residues (Oroudjev et al., 2002). Adhesion between two adjacent cell surfaces of the eukaryote *Dictyostelium discoideum* involves discrete interactions characterized by an unbinding force of about 23 pN. This force probably originates from interactions of individual "contact site A" (csA) glycoprotein molecules (Fig. 8.5; Benoit et al., 2000).

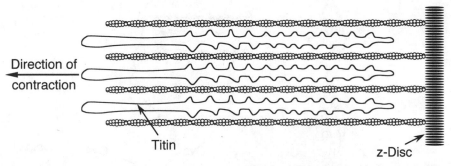

Direction of contraction

Titin

z-Disc

Fig. 8.1 Molecular architecture of skeletal muscle fibers. AFM-related techniques contributed to the understanding of the role of individual titin molecules in such fibers: some skeletal muscle proteins can withstand drags of 600 kp cm^{-2} (see Figs. 8.2 and 8.3; Rief et al., 1997; Gaub and Fernandez, 1998; Carrion-Vasquez et al., 1999)

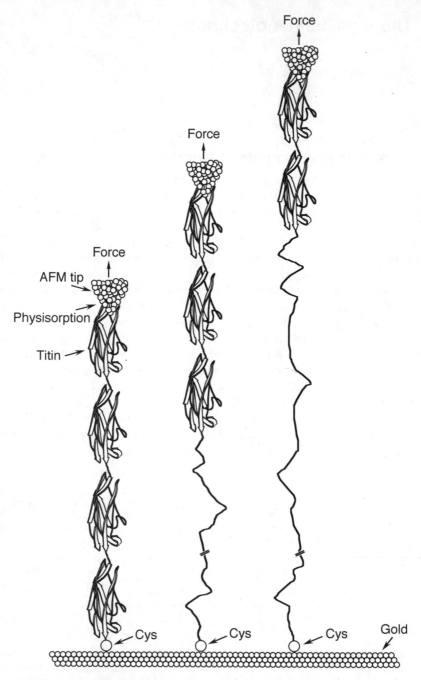

Fig. 8.2 Unfolding of a titin fragment with the help of an AFM (Gaub and Fernandez, 1998; Carrion-Vasquez et al., 1999). The unfolding force for the protein, anchored with a cysteine (Cys) to a gold surface, ranged from about 100 to 300 pN (see Fig. 8.3)

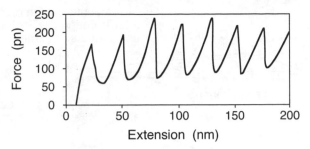

Fig. 8.3 Sketch of a force-distance curve for the unfolding of a titin fragment with the help of an AFM (see Fig. 8.2). The saw tooth-shaped force-extension curve reflects the unfolding of individual titin domains according to an all-or-non mechanism (Gaub and Fernandez, 1998; Carrion-Vasquez et al., 1999)

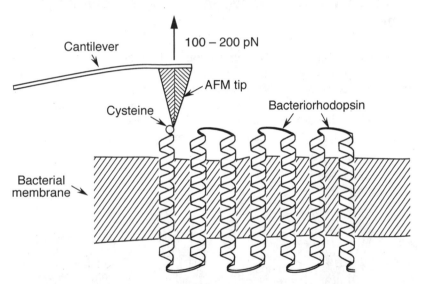

Fig. 8.4 Unfolding of individual bacteriorhodopsins. A force of 100–200 pN is required to remove a bacteriorhodopsin helix from the bacterial membrane (Oesterhelt et al., 2000)

Recoverin, a calcium-myristoyl switch protein, binds to a phospholipid bilayer in the presence of Ca^{2+} with an adhesion force of 48 ± 5 pN (Desmeules et al., 2002). Single molecules of *holo*-calmodulin (i.e. the calcium-loaded form) require a significantly larger force of unfolding by an AFM tip than single molecules of the *apo*-form (Hertadi and Ikai, 2002). Single molecules of the giant filamentous protein titin exhibit mechanical fatigue when exposed to repeated stretch and release cycles (Kellermayer et al., 2001). For further AFM studies on single protein molecules see also Sects. 7.1, 8.2, and 8.3.

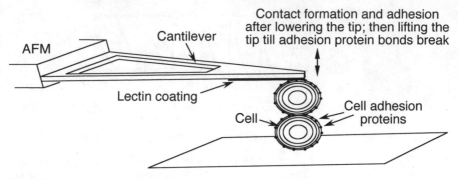

Fig. 8.5 Measurement of discrete interactions in cell adhesion (Benoit et al., 2000)

8.2 Force measurements in a single polymerase-DNA complex

DNA polymerases catalyze DNA replication. The replication reaction requires single-stranded DNA (ssDNA) as a template. In the course of the reaction, a complementary strand of ssDNA is synthesized to the original ssDNA. Already during the polymerization reaction, both strands coil around each other, leading to a shortening of the end-to-end distance of the DNA. Exerting strain on the DNA strand during polymerization can stop and even revert the extension reaction

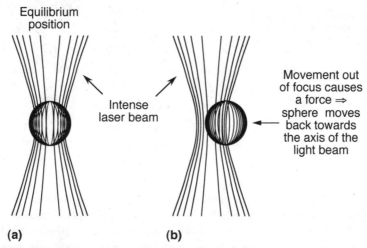

Fig. 8.6 Optical tweezers for the measurement of the effect of template tension on T7 polymerase activity (Fig. 8.7; Smith et al., 1996; Wuite et al., 2000). Regarding the method of optical tweezers see also the groundbreaking studies of single-molecule mechanics by Florin et al. (Florin et al., 1997; Jeney et al., 2001; Pralle and Florin, 2002) and Smith et al. (Bustamante et al., 2000; Liphardt et al., 2001; Smith et al., 2001)

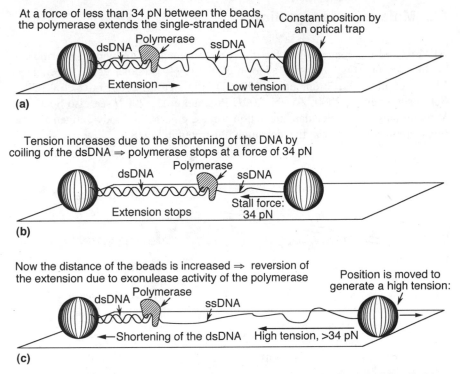

At a force of less than 34 pN between the beads,
the polymerase extends the single-stranded DNA

Constant position by
an optical trap

Polymerase

dsDNA   ssDNA

Extension→   Low tension

(a)

Tension increases due to the shortening of the DNA by
coiling of the dsDNA ⇒ polymerase stops at a force of 34 pN

Polymerase

dsDNA   ssDNA

Extension stops   Stall force:
34 pN

(b)

Now the distance of the beads is increased ⇒ reversion of
the extension due to exonulease activity of the polymerase

Position is moved to
generate a high tension:

Polymerase

dsDNA   ssDNA

←Shortening of the dsDNA   High tension, >34 pN

(c)

Fig. 8.7 Measurement of the effect of template tension on T7 polymerase activity (Smith et al., 1996; Wuite et al., 2000): the polymerase which catalyses DNA replication can work against a maximum force of about 34 pN. Exonuclease activity increases about 100-fold above 40 pN template tension.

(Figs. 8.6 and 8.7; Smith et al., 1996; Wuite et al., 2000). For the measurement of the small forces in the single DNA-protein complex, an optical trap was used: a small bead with DNA attached to it is held into position and moved by an intense laser beam. The main mechanism of the action of such optical tweezers is commonly as follows: The bending of light rays through the refractive sphere is connected with a change of momentum of the light which exerts a force back on the sphere. When the sphere is out of focus of the light beam, these light deflection forces pull the sphere back into focus. Another mechanism is as follows: When an isotropically scattering bead moves out of focus, the momentum of the photons scattered in the direction of the movement increases due to the Doppler effect which decelerates the bead. The larger the light intensity the larger is the deceleration of the movement. Thus, the Brownian motion out of focus is energetically unfavorable relative to the motion into focus. Choosing a wavelength just below an absorption maximum of the bead increases the trapping force since then movement causes increased absorption due to the Doppler-shift of the wavelength and thus an additional momentum slowing down the bead.

8.3 Molecular recognition

AFM-related techniques allow the direct measurement of individual intermolecular interactions (Figs. 8.8–8.11; Florin et al., 1994; Dammer et al., 1995, 1996; Merkel et al., 1999; Strunz et al., 1999; De Paris et al., 2000; Fritz et al., 2000; Schwesinger et al., 2000; Zocchi, 2001; Prechtel et al., 2002; see also Sect. 7.1). Virtually any intermolecular interaction forces, e.g., of antibody-antigen interactions, are measurable with the technique illustrated in Fig. 8.8.

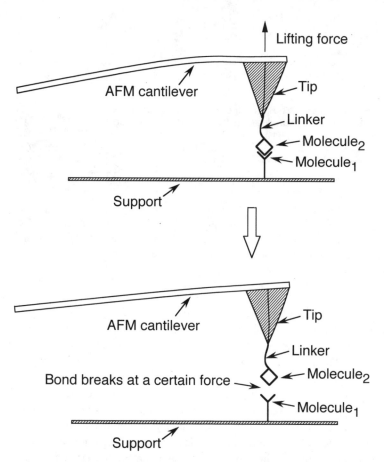

Fig. 8.8 Direct measurement of intermolecular interactions by an AFM-related technique. One of the interacting molecules is immobilized on the surface of the support, the other is connected to the AFM tip by a linker. The tip is approached to the surface so that a specific interaction can take place. Retracting the cantilever ruptures the biophysical interaction. The strength of the interaction is determined from the retract force distance curves (De Paris et al., 2000; Schwesinger et al., 2000; see also Sect. 7.1)

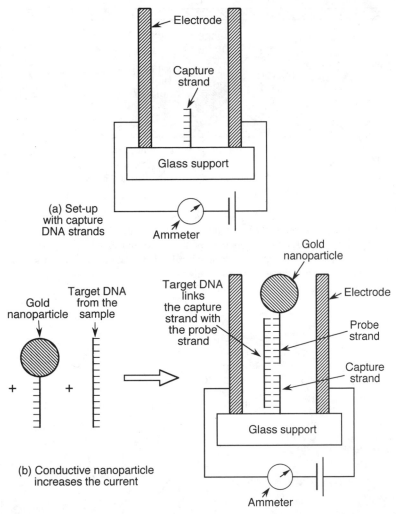

Fig. 8.9 Sensor of biological agents using recognition between single DNA molecules (Park et al., 2002; Service, 2002; see also Demers et al., 2002). (**a**) Two electrodes and single-stranded capture DNA strands are attached to a glass substrate. The capture DNA is complementary to one end of the target DNA of the agent. (**b**) Target DNA and probe strand DNA was added. The probe strand DNA has a gold nanoparticle attached and is complementary to the other end of the target DNA. When all three strands of DNA hybridize together, the gold nanoparticle gets held between the two electrodes. This is detected by an increase of current

Fig. 8.9 illustrates a new type of DNA sensor with potential application for detection of biological contaminants (Park et al., 2002; Service, 2002). It is based on the change of electrical conductivity when gold particles attached to DNA bind

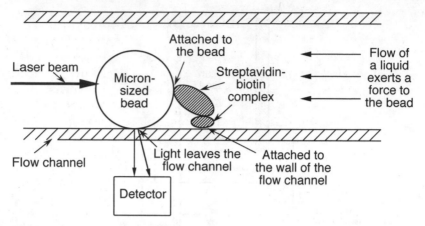

Fig. 8.10 Force measurement on single molecular contacts through evanescent wave microscopy (Zocchi, 2001). The motion of the bead attached to the wall of the flow channel through a single streptavidin-biotin complex is tracked by detecting the evanescent wave as a force is exerted through a flow. This technique allows the direct measurement of the bond rupture force of the molecular complex

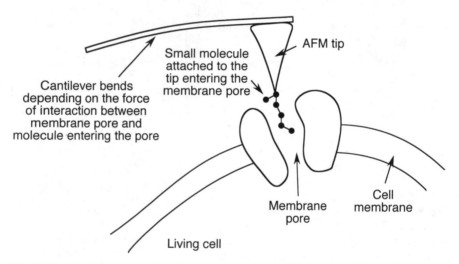

Fig. 8.11 Nanobiosensor using the cantilever of an AFM (Pereira, 2001). Entering or exiting of specific molecules, including medications, from living cells is observed in real-time and the force of interaction measured

to the targeted sample DNA immobilized between two electrodes. A problem of this detection method might be the difficulty to find pieces of DNA that are unique for the organism of interest. In particular, some genetically engineered bacteria and viruses might remain undetected.

8.4 Protein nanoarrays and protein engineering

Lee et al. (2002b) manufactured protein nanoarrays by means of dip-pen nano-lithography (Figs. 8.12 and 8.13): A gold thin-film substrate was patterned, by using an AFM, with a protein-binding chemical linker in the form of dots or grids. Non-patterned surface was inactivated and then protein bound to the linker patterns. Another method for the manufacture of protein nanoarrays is to use germanium pyramids as a support (Fig. 8.14). Calvo et al. (2002) report on the molecular wiring efficiency of glucose oxidase in organized self-assembled nanostructures. Wired protein molecules may be important for future nano-technological tools (Figs. 8.15 and 8.16). Also the design of non-native macro-molecular assemblies, e.g., hexameric helical barrels (Ghirlanda et al., 2002) is an endeavor with implications for nanotechnology (Fig. 8.17). Finally, Fig. 8.18 displays the manufacture of ordered inorganic nanocrystals on top of an array of genetically engineered viruses (Lee et al., 2002c).

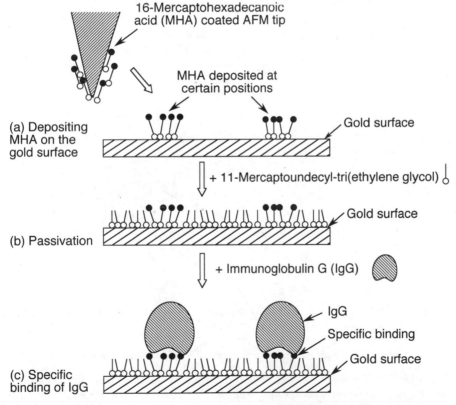

Fig. 8.12 Manufacture of protein nanoarrays for the investigation of molecular interaction and other recognition processes (Lee et al., 2002b; see also Hodneland et al., 2002)

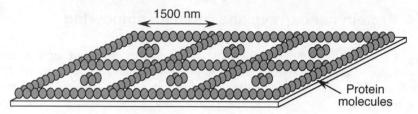

Fig. 8.13 Example of an engineered protein nanoarray (Lee et al., 2002b). Protein arrays with 100- to 350-nm features were fabricated with dip-pen nanolithography (see Fig. 8.12)

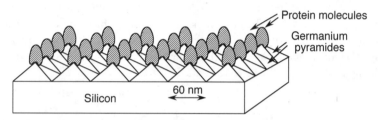

Fig. 8.14 Protein nanoarray manufactured by self-assembly on nanometer-sized germanium pyramids (Riedel et al., 2001). Dynamic contact angle measurements with water droplets revealed that the germanium substrate is highly hydrophilic, and thus should be suitable for adsorption of hydrophilic proteins. Some protein inactivation was observed, however

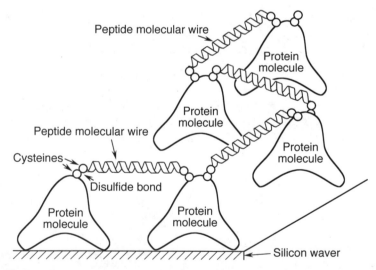

Fig. 8.15 Wired protein molecules might be components in future nanobiotechnological devices (see, e.g., Service, 2001; Calvo et al., 2002; Seeman and Belcher, 2002). Suitable wires are made, e.g., from chemical compounds, peptides, or carbon nanotubes. Proteins may act in such bioelectronic structures, e.g., as redox relays (Calvo et al., 2002)

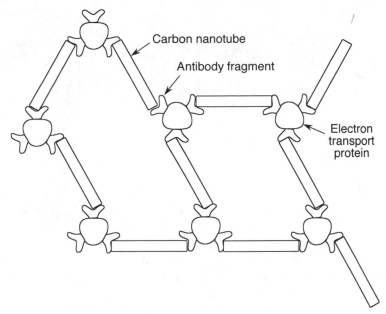

Fig. 8.16 Network of single-walled carbon nanotubes and genetically engineered proteins in a future nanobiotechnological device. Carbon nanotubes display an exceptionally low resistance and are seen as especially useful for the production of bridging nanowires and other nanostructures (see, e.g., Fagas et al., 2002; Odom et al., 2002)

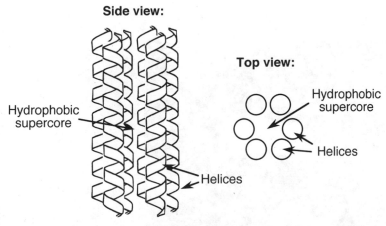

Fig. 8.17 Engineered hexameric helical barrels (Ghirlanda et al., 2002). A dimeric three-helix bundle was designed from first principles. In order to probe the requirements for stabilizing the hexamer, Ghirlanda et al. systematically varied polarity and steric bulk of the residues in the supercore of the hexamer. Formation of the hexameric assembly was best stabilized by changing three polar residues per three-helix bundle to hydrophobic residues (two phenylalanines and one tryptophan)

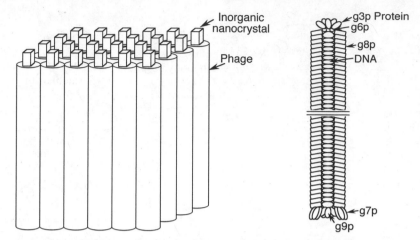

Fig. 8.18 *Left:* ordering of inorganic nanocrystals using genetically engineered phages (Lee et al., 2002c). The bacteriophages form the basis of the self-ordering system. Genetic engineering enables the phages to specifically bind to nanocrystals. *Right:* structure of a phage

8.5 Study and manipulation of protein crystal growth

The fabrication of properly diffracting protein and virus crystals is often the main obstacle for the high resolution of proteins using X-ray methods (see Sect. 4.1).

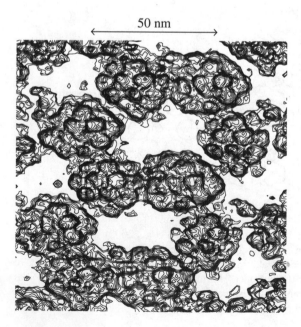

Fig. 8.19 AFM image of the surface of a turnip yellow mosaic virus crystal (Malkin et al., 1995, 2002; Kuznetsov et al., 2000; McPherson et al., 2000, 2001). AFM investigation revealed the sources of crystal disorder and mechanisms of their formation

AFM is exquisitely useful for the study and manipulation of crystal growth (Fig. 8.19; Durbin and Carlson, 1992; Durbin et al., 1993; Malkin et al., 1995; McPherson et al., 2000; Mollica et al., 2001; Biscarini et al., 2002). Most protein and virus crystals grow, through a process of two-dimensional nucleation, by formation of new crystal layers (McPherson et al., 2000). Scratching the surface of a lysozyme crystal which was completely covered by an impurity stopping crystal growth resulted in resumption of crystal growth (McPherson et al., 2000).

8.6   Nanopipettes, molecular diodes, self-assembled nanotransistors, nanoparticle-mediated transfection and further biophysical nanotechnologies

Bone cells respond to stretching by an AFM tip with activation of stretch-activated ion channels (Charras and Horton, 2002). Dissecting bacterial surface layers with an AFM tip provided a better understanding of the high stability of this protective bacterial surface coat (Fig. 8.20; Scheuring et al., 2002). Micronized salbutamol particles stick to glass stronger than to polytetrafluoroethylene (Eve et al., 2002). For further nanobiotechnological innovations see Figs. 8.21–8.27 and Sect. 7.1.

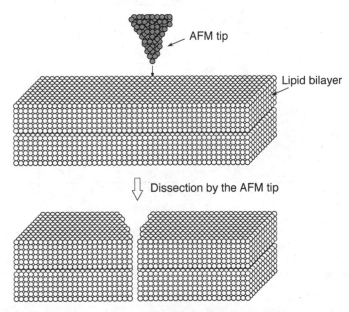

Fig. 8.20 Unzipping a double layer of lipids adsorbed to mica (not shown), with an AFM tip (Scheuring et al., 2002). Using the AFM stylus as a nanodissector, native bacterial surface layers were separated and their mechanical and protective properties against hostile environments examined

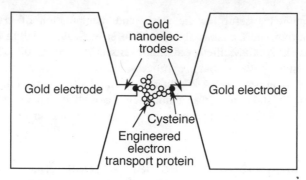

Fig. 8.21 Conduction between nanoelectrodes through a single molecule (Reimers et al., 2002) or a few organic molecules forming a nanocrystal (Rinaldi et al., 2002). In some cases, rectifying behavior is observed, e.g., for deoxyguanosine nanocrystals (Rinaldi et al., 2002)

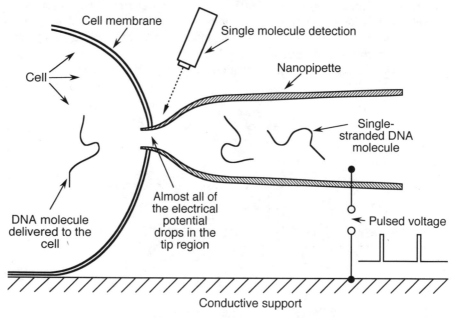

Fig. 8.22 Programmable delivery of DNA through a nanopipette (Ying et al., 2002; Bruckbauer et al., 2002b). The conical geometry of the pipette causes most of the electrical potential drop to occur in the tip region. Pulsatile delivery of DNA molecules is achieved by controlling the applied voltage

The nanopipette in Fig. 8.22 was designed for controlled delivery of macromolecules into living cells (Ying et al., 2002). A voltage applied to the nanopipette moves the DNA molecules slowly towards the tip of the pipette. Since, because of the conical geometry of the nanopipette, most of the potential drop occurs in the tip region, the DNA molecules are rapidly delivered to the cell once they have reached the tip region. In combination with single molecule detection, individual DNA molecules can be delivered to the living cell in a controlled manner.

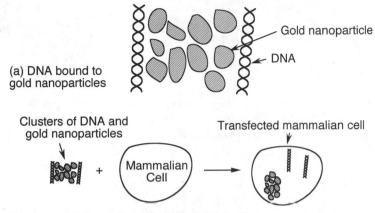

Fig. 8.23 Gold nanoparticle-mediated transfection of cells (Sandhu et al., 2002). (a) Mixed monolayer protected gold cluster functionalized with quaternary ammonium chains binding to DNA. (b) Schematic of the transfection process

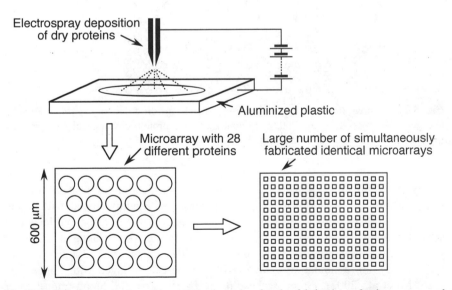

Fig. 8.24 Electrospray deposition of dry proteins for the fabrication of microarrays and nanoarrays (Avseenko et al., 2001, 2002). These arrays can detect antibodies in plasma samples from mice immunized with the proteins used for the arrays

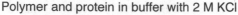

Polymer and protein in buffer with 2 M KCl

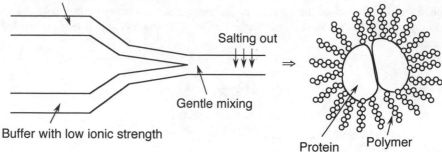

Fig. 8.25 Continuous-flow preparation of 100-nm nanoparticles for drug delivery, protein delivery, and gene therapy (Prokop et al., 2001; Davda and Labhasetwar, 2002; Igartua et al., 2002; Konan et al., 2002; Haas and Lehr, 2002)

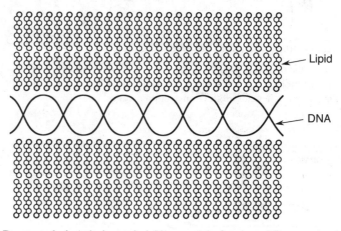

Fig. 8.26 De-novo designed virus-mimicking particle for drug delivery, protein delivery, and gene therapy (Xu et al., 2002). In contrast to the protein envelope of most viruses, this particle has an envelope made from lipids

Fig. 8.25 illustrates a technique for the fast production of protein-polymer nanoparticles: A solution of polymer and protein is gently mixed with a salt solution. The salting-out effect causes the self-organization of nanoparticles that have different surface properties than the protein in the core (Prokop et al., 2001; Davda and Labhasetwar, 2002; Igartua et al., 2002; Konan et al., 2002; Haas and Lehr, 2002). In a similar way, also DNA can be camouflaged with lipids (Fig. 8.26; Xu et al., 2002). Some of these particles can cross the blood-brain barrier and are seen as promising candidates for future gene therapy and drug delivery.

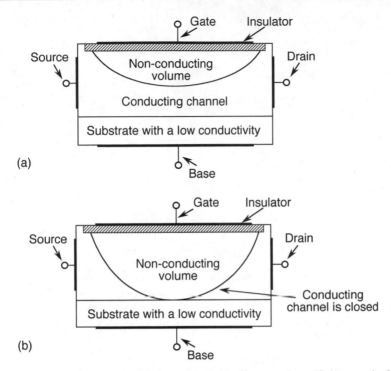

Fig. 8.27 Self-assembled monolayer organic field-effect transistor (Schön et al., 2001a, 2001b). The figure depicts an example of the principle of operation of a field-effect transistor: The source-drain conductivity is controlled through the gate which is electrically insulated from the device itself: the width of a conducting channel between source and drain is varied by adjusting the voltage between gate and base. **(a)** Open channel between source and drain: current can flow between source and drain. **(b)** The source-drain channel is closed by application of a voltage between gate and base: now the source is isolated from the drain

A field effect transistor was manufactured by using self-assembly of organic molecules (Schön et al., 2001a, 2001b). For an example of the principle of operation of field-effect transistors see Fig. 8.27.

9 Proteomics: high throughput protein functional analysis

Currently, there is a major effort, on a genome-wide scale, to map protein-drug interactions and to discover drug targets (Sect. 9.1), to map protein-protein interactions (Sect. 9.2), to discover chemical activity of proteins (Sect. 9.3), and to resolve protein structures (Sect. 9.5). This effort, called proteomics, provides significant knowledge of the biology of organisms far beyond the level of sequence information (see, e.g., Adam et al., 2002b; Burbaum and Tobal; Edwards et al., 2000, 2002; Christendat et al., 2000; Figeys, 2002a, 2002c; Gallardo et al., 2002; Hubbard, 2002; Kersten et al., 2002; Koshland and Hamadani, 2002; Lin and Cornish, 2002; Liu et al., 2002; Morrison et al., 2002; Natsume et al., 2002; Yarmush and Jayaraman). The system-wide study of proteins and as well non-proteinaceous interaction partners largely employs protein microarray technology (see, e.g., MacBeath, 2002; Gera et al., 2002; Kukar et al., 2002; Talapatra et al., 2002) and bioinformatic methods (see, e.g., Bork, 2002).

Proteomics-based approaches for the study of organ-specific regulatory and signaling cascades are seen as a key for a better understanding and therapeutical management of diseases (e.g., Jäger et al., 2002). Proteomics has provided new vaccine candidate antigens (Klade, 2002; Nilsson, 2002; Vytvytska et al., 2002). The identification of individual proteins abnormally expressed in tumors may have an important relevance for making diagnosis, prognosis, and treatment (e.g., Celis et al., 2002; Dwek and Rawlings, 2002; Jain, 2002; Michener et al. 2002; Zheng et al., 2003). Proteomics analysis of the neurodegeneration in the brain of transgenic mice discovered 34 proteins with significantly changed intensity (Tilleman et al., 2002). A proteomics approach was used to identify the translation products of squid optic lobe synaptosomes (Jimenez et al., 2002). A central nervous system (CNS) proteome database derived from human tissues is expected to significantly accelerate the development of more specific diagnostic and prognostic disease markers as well as new selective therapeutics for CNS disorders (Rohlff and Southan, 2002). Proteomics provides an extremely powerful tool for the study of variations in protein expression between different ages and for the understanding the changes that occur in individuals as they become older (Cobon et al., 2002).

Innovations towards higher throughput and cost cutting include mass spectrometry advances (Sects. 9.1 and 9.2), DNA microchips (Sect. 9.1), protein microchips (Sect. 9.2), genetic hybrid systems (Sect. 9.2), and lab-on-a-chip technology (Sect. 9.4).

9.1 Target discovery

Two-dimensional electrophoresis and mass spectrometry (Fig. 9.1) are widely used for the study of protein composition and protein changes in humans, animals, and plants. Important applications are (a) the identification of biomarkers specific for certain cell types, disease states, or aging processes, and (b) the study of protein composition changes as a response to drug treatment.

Also, high throughput microarray-based assays hold tremendous promise for the discovery of proteins connected with diseases (Fig. 9.2).

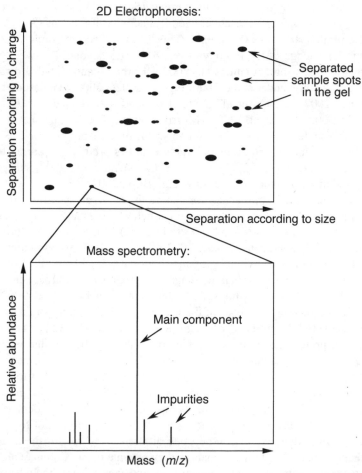

Fig. 9.1 Discovery of proteins relevant to a certain disease by two-dimensional poly-acrylamide gel electrophoresis and mass spectrometry (see, e.g., Edwards et al., 2000; Blomberg, 2002; Kersten et al., 2002; Man et al., 2002; Mo and Karger, 2002; Rohlff and Southan, 2002)

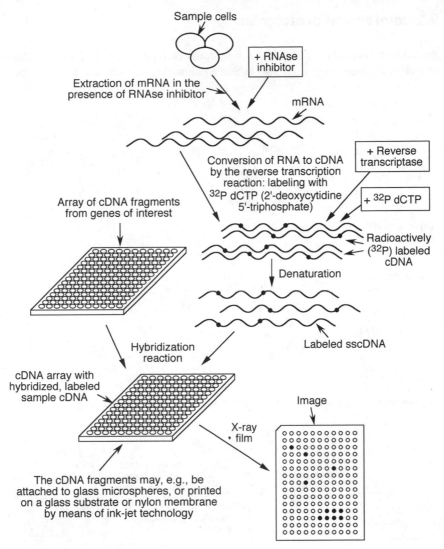

Fig. 9.2 Discovery of proteins relevant to a certain disease, e.g., cancer markers, by detection of changes in the abundance of mRNA by means of cDNA (complementary DNA) microarray technology (supplied, e.g., by SuperArray, Inc., Bethesda, MD). cDNA chips with spot sizes of 10–500 μm are commonly fabricated by high speed robotics or ink-jet printing on glass or nylon substrates. Every spot contains a different, 100–10,000 bases long, immobilized probe cDNA fragment which is complementary to targeted cDNA. The targeted, radioactively labeled cDNA is synthesized by reverse transcriptase from mRNA of the sample cells. Single stranded target cDNA is hybridized with complementary cDNA of the array and non-binding cDNA rinsed off with buffer. Detection of the pattern of radioactivity of the array then shows which mRNA was present in the sample cells, and thus which proteins were expressed

9.2 Interaction proteomics

Analysis of several 100,000 protein-protein interactions using microarray technology (Fig. 9.3) and the yeast two-hybrid system (Fig. 9.4) has led to dozens of

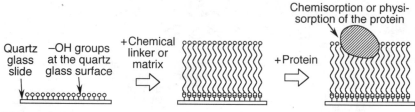

(a) Preparation of a single spot of the protein microarray

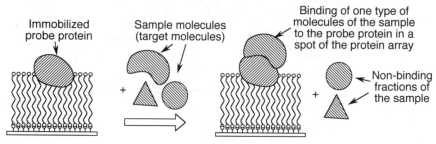

(b) Measurement of protein-protein interactions with the spot of the array

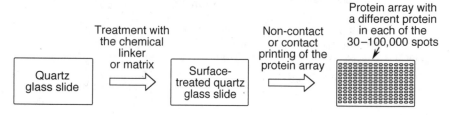

(c) Manufacture of the protein microarray

Fig. 9.3 **(a)** Manufacture of a spot of a protein microarray for the assay of protein-protein interactions (see, e.g., Grayhack and Phizicky, 2001; MicroSurfaces, Inc., Minneapolis, MN). For a better maintenance of structural integrity, protein immobilization is carried out on a matrix or layer of a chemical linkers that provide a native-like environment for embedded proteins. **(b)** Sample protein molecules (target molecules) interact with a probe molecule of a spot of the array. Non-binding sample proteins are simply washed off with buffer. Sample molecules are radioactively or fluorescence labeled so that binding of target protein molecules with probe protein molecules can be detected. **(c)** Manufacture of an array with 30–100,000 individual probe proteins immobilized on a single slide by chemical treatment of the surface of a quartz glass slide and ink-jet or contact printing of the protein spots

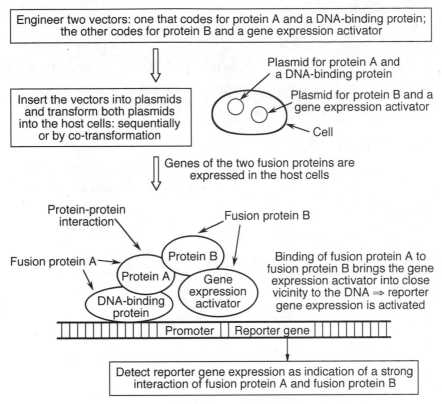

Fig. 9.4 Discovery of protein-protein interactions by means of the yeast two-hybrid system (McCraith et al., 2000; Ito et al., 2002; Stagljar and Fields, 2002): Two vectors are constructed so that (i) one contains the code for a protein (protein A: e.g., a predicted open reading frame) followed by the code for a DNA-binding protein, and (ii) the other contains the code for another protein (protein B: e.g., another predicted open reading frame) followed by the code for a gene expression activator. Expression of both vectors in the cell yields a DNA-binding fusion protein and an expression-activating fusion protein. Attraction of both fusion proteins brings the expression activator into vicinity to the DNA. This leads to reporter gene activation. Diploids expressing the two-hybrid reporter gene in the host cells are then identified

novel findings of important intermolecular interaction (see, e.g., McCraith et al., 2000; Ito et al., 2002; Stagljar and Fields, 2002).

Figs. 9.5 and 9.6 present a mass-spectrometric method for the analysis of a large number of protein-protein and protein-drug interactions, respectively, without need of 2D-chromatography or electrophoresis. Possible ambiguities in the assignment of mass peaks may be resolved by the technique of ion fragmentation (see, e.g., Fig. 3.27). The discovery of new drug targets is highly important for developing new drugs (e.g., Pillutla et al., 2002; Whitelegge and le Coutre, 2002).

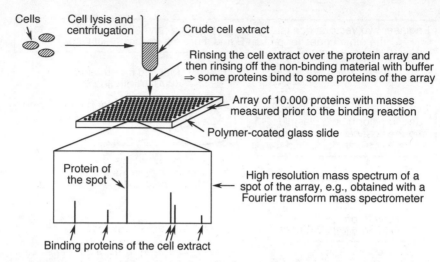

Fig. 9.5 Analysis of millions of protein-protein interactions in one organism without 2D-chromatography or electrophoresis: Target proteins immobilized on the microarray interact simultaneously with all proteins expressed in a certain type of cells. After rinsing off the non-binding molecules, a high resolution mass spectrogram is recorded for each spot of the array. Each mass spectrogram shows a peak corresponding to the target protein and possibly further peaks corresponding to binding proteins. The binding proteins are then identified by their masses

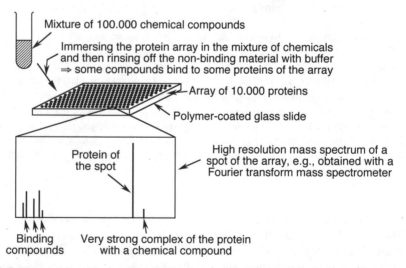

Fig. 9.6 Large-scale analysis of interactions of chemical compounds with proteins without 2D-chromatography or electrophoresis: The protein array is dipped into the mixture of chemicals. After some time, non-binding chemicals are washed away with buffer. The spots are then mass-spectrometrically analyzed. Since the masses of the involved proteins and chemicals were previously measured, binding chemicals can easily be identified

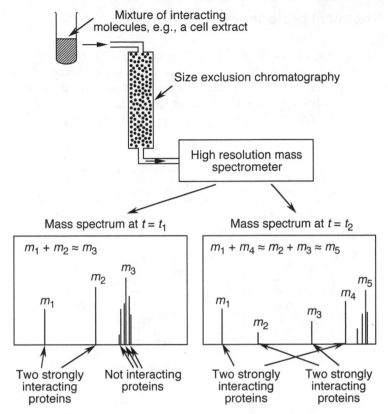

Fig. 9.7 Direct determination of strong protein-protein interactions in a cell extract without need for the manufacture of arrays. First the mixture of monomeric proteins and protein-protein complexes is separated according to size by gel chromatography. Conditions for the chromatography are chosen such that strong complexes do not completely dissociate. Then mass spectrometry is performed for each chromatographic fraction with identifies the dimeric complexes and their two interacting macromolecules in the fraction: During ionization in the mass spectrometer, the complexes dissociate causing two peaks in the spectrum. These twin peaks are identified by their total mass which is roughly equal to the mass of the non-binding macromolecules. This method may analogously be applied to map out other strong macromolecular interactions

High resolution mass spectrometry even allows the simultaneous mapping-out of a large number of protein-protein and protein-drug interactions without use of microarrays (Fig. 9.7). A fast size exclusion chromatography separates the complicated mixture of interacting molecules into fractions. The parameters for chromatography were chosen such that strongly interacting molecule complexes remain together. Chromatographic fractions are then mass-spectrometrically analyzed. Strongly interacting dimers of molecules appear in the spectrogram as pairs with a total mass of about that of the monomers in the fraction.

9.3 Chemical proteomics

Chemical proteomics assigns molecular and cellular functions to thousands of identified or predicted gene products. Assaying activities of large pools of constructed strains with subsequent deconvolution of active pools is an efficient method to discover new functions of genes (see, e.g., Fig. 9.8; Martzen et al., 1999; Grayhack and Phizicky, 2001; Adam et al., 2002a; Phizicky et al., 2002).

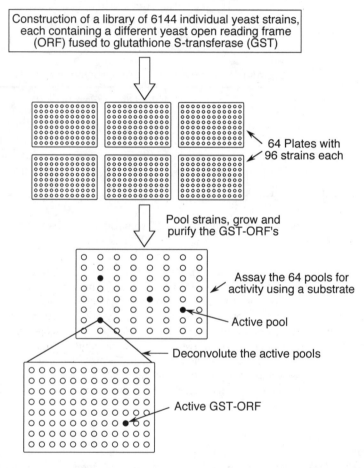

Fig. 9.8 Example of chemical proteomics (Martzen et al., 1999; Grayhack and Phizicky, 2001; Phizicky et al., 2002). A library of 6144 yeast strains was constructed. Each strain expresses a unique yeast open reading frame (ORF) as a GST-ORF fusion (GST, glutathione S-transferase). Each 96 strains were pooled and biochemical activity of the pools was assayed. Active pools were deconvoluted using the library of strains to identify the GST-ORF responsible for activity. Several previously unknown biochemically active gene products were discovered

9.4   Lab-on-a-chip technology and mass-spectrometric array scanners

Protein and DNA microarrays are increasingly often processed with the lab-on-a-chip technology (Figs. 9.9 and 9.10): tiny channels etched into a glass slide, microswitches, micromixers, and other small devices act as small chemical factories.

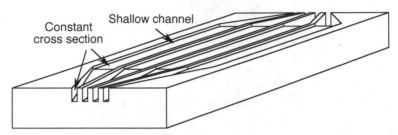

Fig. 9.9 Simplified example for lab-on-a-chip technology (see, e.g., Swedberg et al., 1996; Swedberg and Brennen, 2001; Cheng et al., 2002; Figeys 2002b, Laurell and Marko-Varga, 2002). Tiny channels are microfabricated or etched into the support. The top is then sealed with another plate (not shown). A single chip may contain all the channels, switches and reservoirs necessary for complicated multi-stage chemical reactions

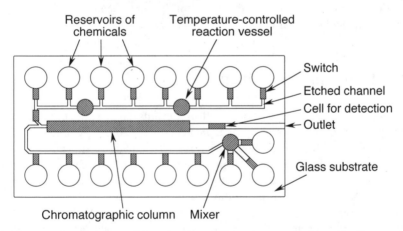

Fig. 9.10 Example of lab-on-a-chip technology. Channels are etched into the glass support

Scanning of protein and DNA arrays with fluorescence detectors usually requires special labeling of the sample and may be prone to errors due to limitations of sensitivity and due to unspecific binding. More importantly, mixtures of signals can often not be resolved. Automatic mass-spectrometric

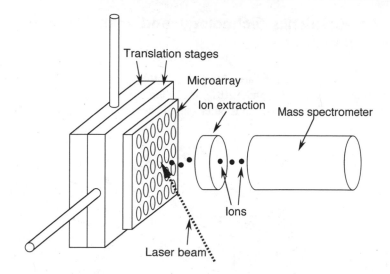

Fig. 9.11 Mass-spectrometric array scanner for automatic mass-spectrometric measurement of protein and DNA microarrays. Step motor controlled translation stages rapidly move the microarray into position. A mass spectrum for each of the spots is automatically taken and analyzed. The scanner can be used, e.g., in the methods outlined in Figs. 9.5 and 9.6

detection of the molecules in the spots of the array can greatly enhance the information yield (Fig. 9.11).

9.5 Structural proteomics

Proteomics is driving a substantial effort towards large-scale protein structure prediction (see, e.g., Renfrey and Featherstone, 2002; Schmid, 2002) and determination. High resolution structure determination still relies on X-ray crystallography and NMR (nuclear magnetic resonance). Since both methods are expensive and time-consuming, further optimization of the methods is being in progress. NMR peaks can now automatically be assigned to the corresponding amino acid residues (Nilges et al., 1997; Heinemann et al., 2001). The Berlin Protein Structure Factory develops and applies large-scale NMR and crystallographic methods (Heinemann et al., 2000, 2001; Boettner et al., 2002).

On the other hand, structure computer simulation using simplifications of the conformational space of proteins is rapidly progressing (see Sect. 1.4), and so it can be hoped that comparably inexpensive computational methods will make an increasing contribution to structural proteomics in the near future.

10 Ion mobility spectrometry

10.1 General design of spectrometers

Ion mobility spectrometry was developed for the simple and cheap detection and characterization of organic compounds (Cohen and Karasek, 1970; Karasek, 1970; Caroll et al., 1971; Caroll, 1972; Cohen et al., 1972; Cohen and Crowe, 1973; Vora et al., 1987; St. Louis and Hill, 1990; Campbell et al., 1991; Burke, 1992; Eiceman and Karas, 1994; Taylor, 1996; Baumbach and Stach, 1998; Baumbach and Eiceman, 1999; Saurina and Hernandez-Cassou, 1999; Asbury and Hill, 2000; Purves et al., 2000; Wu et al., 2000; Beegle et al., 2001; Eiceman et al., 2001; Matz and Hill, 2001; Stone et al., 2001). The ion mobility spectrum reflects the ion mobilities which correlate well with the size-to-charge ratios of the sample compounds.

In the ion mobility spectrometer (IMS), (a) sample molecules in the vapor phase are ionized, (b) the charged sample molecules (ions) are accelerated by an electric field, and (c) their time of flight in the gaseous medium of the drift channel is measured and recorded (Figs. 10.1–10.6). These simple spectrometers can detect and analyze astonishing tiny traces of small and as well large molecules and clusters.

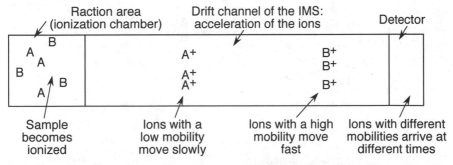

Fig. 10.1 Principle of operation of ion mobility spectrometers: Molecular ions are generated and accumulated in the reaction area. A gating pulse transfers the molecular ions to the drift channel where said ions are accelerated by an electric field. Ions with different mobilities in the gaseous medium of the drift channel arrive at different times at the detector

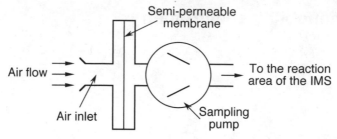

Fig. 10.2 Design of a sample inlet for an IMS. The sampling pump draws air through the semi-permeable membrane which attenuates large dust particles and other interferents (see also Spangler, 1982). A suitable membrane is, e.g., a 5–50 μm polytetrafluoroethylene (PTFE) foil or silicone rubber membrane (Spangler and Carrico, 1983; Kotiaho et al., 1995). For IMS for detection of biological agents, a metal grid instead of a membrane may be more appropriate because of its better transparency for high molecular-weight compounds

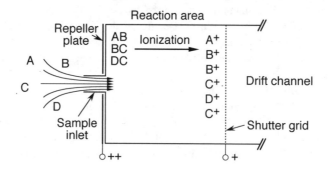

(a) Positive ion mode

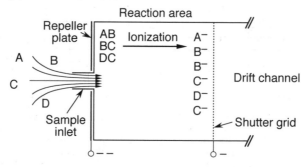

(b) Negative ion mode

Fig. 10.3 Reactions of the sample in the reaction area. Sample molecules are ionized, either directly by dissociation or indirectly by clustering with other ions. In the positive ion mode, positive ions are repelled from the repeller plate and accumulated in front of the shutter grid. In the negative ion mode, negative ions are analogously accumulated. Many IMS operate alternatingly in the positive and negative ion mode

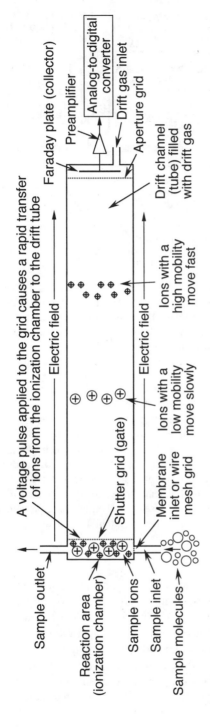

Fig. 10.4 Principle of operation of an ion mobility spectrometer (IMS) (Cohen and Karasek, 1970; Karasek, 1970; Caroll et al., 1971; Keller, 1975; Eiceman and Karas, 1994). Sample molecules are injected into the reaction area (ionization chamber) and ionized. A thin membrane or a wire mesh grid separates reaction area and drift channel (drift tube, drift region). An electrical pulse applied to the shutter grid (gate, grid, gating grid) transfers the ions into the drift channel where the ions are further accelerated by the electric field which is generated by guard rings (see Figs. 10.5 and 10.6). The time of flight of the ions in the gaseous phase is measured with the help of a Faraday plate (collector plate, Faraday cup) or collector grid. The mobility of a gas phase ion is a measure of its collision cross section which in turn depends on its size and structure. Small ions with compact structures have smaller cross sections than large ions with open structures. Consequently, since different ions have different mobilities in the gas of the drift tube, they can result in distinct peaks in the ion mobility spectrum. A computer (not shown) determines the identity of the sample molecules by matching the spectra to reference signatures. In order to reduce the noise, the IMS is enclosed in a grounded copper foil (not shown). The collector is directly connected with a 10^{10}-V/A preamplifier via a cable of only a few mm length. The feedback resistor of the preamplifier is selected for a low noise level. Since fluctuations of the electric field of the drift channel add noise to the signal of the Faraday plate, the voltage supply for the guard rings is highly stabilized. In contrast to mass spectrometers, this device needs no hot filaments with a limited lifetime, electron multipliers, energy-consuming vacuum pumps, or expensive vacuum tubes, and its sensitivity can be several orders of magnitude higher than that of a mass spectrometer. Drift times of macromolecules are usually milliseconds to seconds in 3–20 cm drift tubes with drift voltage gradients of about 100–1000 V/cm. For most IMS, a few seconds are required after each measurement to purge the drift channel

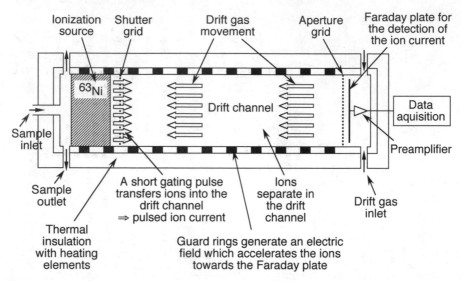

Fig. 10.5 Design of an IMS (Cohen and Karasek, 1970; Karasek, 1970; Caroll et al., 1971; Eiceman and Karas, 1994; Matz and Schröder, 1996, 1997; IUT Institute for Environmental Technologies, Berlin, Germany; Bruker Daltonik, Bremen, Germany; Graseby Dynamics, London, U.K.; Barringer Instruments, Warren, NJ). The ionization source ionizes sample molecules. Guard rings generate an electric field which accelerates the sample ions towards the shutter grid where they accumulate. After a few seconds, a voltage pulse is applied to the shutter grid causing the release of the sample ions into the drift channel. Now the electric field can further move the sample ions towards the ion detector (Faraday plate). Different ions interact differently with the drift gas molecules in the drift channel. This causes the ions to spread out according to their different mobilities. The recorded ion mobility spectrum corresponds to differences in the time of flight of the sample ions. Typically, the drift channel has a length of a few cm, and the electric field strength of in the drift channel is $100–1000$ V cm^{-1}. At this field strength and length, small organic compounds which have mobilities of a few cm^2 V s^{-1} need about $1–100$ milliseconds to reach the detector. Biomacromolecules typically travel about $1–3$ orders of magnitude slower. Drift channel and reaction area are enclosed in a thermal isolation with heating elements. The operating temperature for the detection of biological agents is typically $100–150$ °C

The IMS is comprised of (a) a thermally isolating housing, a source of ionization, reaction area, shutter grid, drift channel with guard rings, possibly an aperture grid, collector, (b) a source of clean gas or a gas filter, (c) a shutter controller, (d) a high voltage supply, (e) an electrometer, (f) temperature control instrumentation, (g) a computer-aided data collection and processing unit.

The major advantage of IMS is the extreme sensitivity (see Sect. 10.2). Another important advantage of IMS, which do not need vacuum parts, is the lower cost and lower energy consumption relative to most mass spectrometers. This makes it particularly suitable for large-scale field applications.

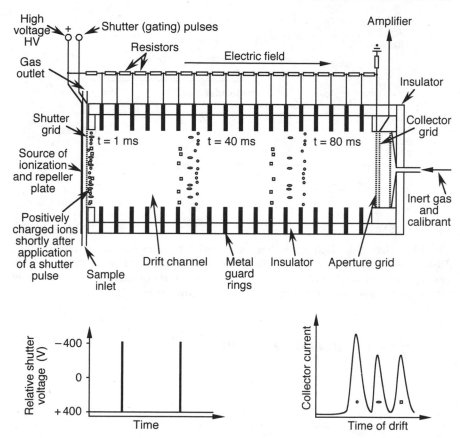

Fig. 10.6 *Top:* design of an IMS containing 18 stacked copper guard rings that are separated by insulating ceramics spacers and connected by resistors. These resistors are selected for low noise. The voltage supply for the guard rings is stabilized to better than 0.1% rms. In order to ensure a homogenous electric field, the guard rings have sufficiently narrow separations. In this example the IMS is operated in the positive ion mode. *Bottom left:* example of the shutter voltage: Most of the time, a positive shutter voltage prevents positively charged ions from entering the drift channel. After a period of time during which positive ions accumulate in front of the shutter grid, a negative shutter voltage pulse is applied. Now positive ions can enter the drift channel where they are further accelerated. *Bottom right:* recorded spectrum

A suitable inert gas for the detection of many organic compounds is nitrogen, but clean air can also serve for this purpose. Cleaning the drift gas from water vapor and sample residues is frequently performed with molecular sieves (Fig. 10.7). These molecular sieves are made from zeolites which are certain aluminosilicates. Zeolites can adsorb water molecules and other small organic compounds. After a few weeks or month of use in an IMS, the zeolite is saturated. It can be reconstituted by heating it in an oven and be re-used.

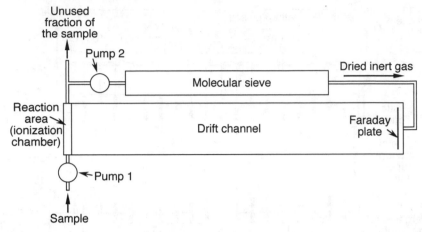

Fig. 10.7 Drying the inert gas with a molecular sieve (Carnahan and Tarassov, 1998; Taylor and Turner, 1999). Suitable dimensions for a length of the drift channel of 5–20 cm, a diameter of the drift channel of 2 – 3 cm, and a height of the ionization chamber of 0.5–1 cm are: flow rate of pump 1: 5–50 ml min^{-1}; flow rate of pump 2: 50–500 ml min^{-1}

10.2 Resolution and sensitivity

The resolution, R, of an IMS is defined as

$$R = 0.5 \cdot t_d \, \tau^{-1} \, , \tag{10.1}$$

where t_d is the drift time of the peak, τ and is its temporal width at half peak height (St. Louis and Hill, 1990).

Important factors affecting the resolution are: (a) initial pulse width and shape, (b) broadening by Coulomb repulsion between the ions in both the reaction region and drift channel, (c) the initial ion pulse width (shutter pulse width), (d) ion-molecule and ion-ion reactions in the reaction region, (e) gate depletion, (f) ion-molecule reactions in the drift region, (g) spatial broadening by diffusion of the ion packet during the drift, (h) temperature and pressure inhomogeneities within the spectrometer, (i) capacitive coupling between aperture grid and collector, and (j) the response time of the preamplifier, amplifier, and analog-to-digital converter. Broadening by Coulomb repulsion is particularly severe in narrow designs of reaction region and drift channel.

R strongly depends on the design of the spectrometer, but can theoretically exceed 1000 (Figs. 10.8–10.10). A high resolution requires a relatively bulky design with a drift channel of sufficient diameter and large length and a high drift voltage. Most commercial IMS have resolutions of only 20–150 since they are mainly optimized for low weight and portability.

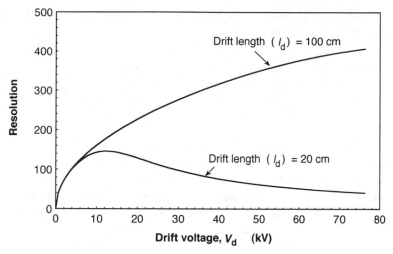

Fig. 10.8 Theoretical resolution of an IMS at different drift lengths. At short drift lengths, increasing the drift voltage above an optimum does not further improve the resolution. The parameters in this example are: temperature, 130 °C; ion mobility, 10^{-5} m^2 V^{-1} s^{-1}; initial ion pulse width, 1 ms; shutter pulse voltage, 1000 V; length of the reaction area, 3 mm; distance between aperture grid and Faraday plate, 1 mm; voltage at the aperture grid, 500 V

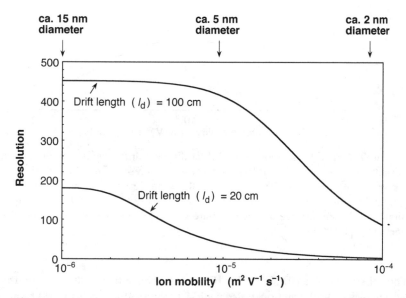

Fig. 10.9 Theoretical resolution of an IMS at different drift lengths and ion mobilities: long shutter pulses enhance the sensitivity, but limit the resolution for small ions. The parameters in this example are: temperature, 130 °C; drift voltage, 80 kV; initial ion pulse width, 1 ms; shutter pulse voltage, 1000 V; length of the reaction area, 3 mm; distance between aperture grid and Faraday plate, 1 mm; voltage at the aperture grid, 500 V

R is approximately given by Eq. 10.2 (St. Louis and Hill, 1990; Hill et al., 1990; Leonhardt et al., 2001):

$$R = \cfrac{1}{\sqrt{16\ln 2\dfrac{k_B T}{qV_d} + \dfrac{K^2 V_d^2}{l_d^4}\left(t_{Pulse} - \dfrac{s^2}{KV_{Pulse}}\right)^2 + \left(\dfrac{a^2 V_d}{l_d^2 V_{ap}}\right)^2 + \dfrac{t_{Pulse}^2 K^2 V_d^2}{l_d^4}}} \qquad (10.2)$$

where k_B, Boltzmann constant $(1.3807 \times 10^{-23}\ \text{J K}^{-1})$; T, absolute temperature; q, ion charge; V_d, drift voltage; l_d, drift length; K, ion mobility; t_{Pulse}, initial ion pulse width; V_{Pulse}, shutter pulse voltage; s, distance between space charge in the reaction area and shutter grid (roughly half the length of the reaction area); a, distance between aperture grid and Faraday plate; V_{ap}, voltage at the aperture grid.

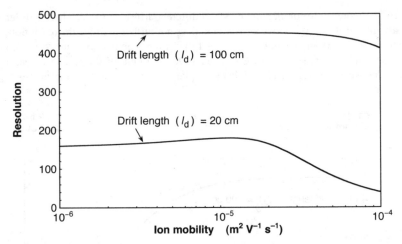

Fig. 10.10 Theoretical resolution of an IMS at different drift lengths and ion mobilities. Short shutter pulses and long drift lengths allow the high resolution of both medium-sized and as well large molecules. The parameters in this example are: temperature, 130 °C; drift voltage, 80 kV; initial ion pulse width, 100 μs; shutter pulse voltage, 1000 V; length of the reaction area, 3 mm; distance between aperture grid and Faraday plate, 1 mm; voltage at the aperture grid, 500 V

The sensitivity of an IMS can be 1000 times grater than that of a good mass spectrometer: with conventional electronics, theoretically down to 1000 ions are still detectable corresponding to 3×10^{-19} g TNT or 2.5×10^{-16} g botulinum toxin ($M_w = 150$ kDa). For comparison, the fingerprint of a person likely contains many billion times more residue of organic compounds. Practical detection limits are typically on the order of 0.1 to 100 ppbv. One reason for the often superior sensitivity relative to mass spectrometers is the simple and large sample inlet. Many IMS contain a thin large area-polymer membrane inlet or a meal grid inlet with cause only little sample losses (Fig. 10.3; Spangler, 1982; Spangler and Carrico, 1983; Kotiaho et al., 1995). In some IMS, the sample inlet solely consists of a tube with a filter which removes dust particles and in some cases water from the sample prior to sample injection into the IMS.

Short gating pulses and high humidity tend to lower the sensitivity. All means of noise reduction, such as (a) small collector capacity, (b) a highly stabilized voltage of the guard rings, (c) an electric shielding of the whole IMS, and (d) low-noise resistors in the electronics, tend to improve the sensitivity. Further, for a high sensitivity and low memory effects, the IMS must be built from materials which adsorb extremely little if any sample molecules, e.g., special ceramics.

Sensitivity and resolution are also affected by the presence or absence of an aperture grid: The function of the aperture grid is to capacitively decouple the collector (Faraday plate or collector grid) from the approaching ion cloud. Without the aperture grid, the collector senses the approaching ion cloud several millimeter prior to its arrival, resulting in line broadening (St. Louis and Hill, 1990). On the other hand, the aperture grid neutralizes ions which decreases the ion current and sensitivity by a factor of roughly 3.

10.3 IMS-based "sniffers"

IMS sniffers (Figs. 10.11 and 10.12) are hand-carried IMS-based devices, mainly manufactured for the detection of dangerous substances. They contain a small IMS and a database of the signatures of substances of interests and interferents. After a few seconds of measurement, the processing unit identifies and signalizes detected agents. The sniffers weighting about 2–10 kg are roughly 25–40 cm long. False alarm rates for the detection of chemicals are quite often 0.01–1%.

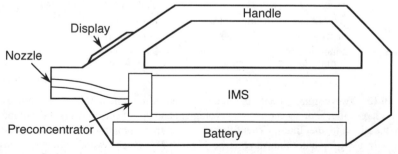

Fig. 10.11 Hand-carried "sniffers" comprising a chemical preconcentrator and an IMS (made, e.g., by Sandia Corporation, Albuquerque, NM, and Barringer Instruments, Inc., Warren, NJ)

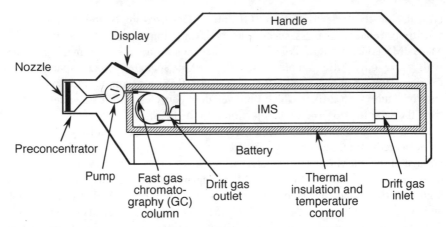

Fig. 10.12   Hand-carried sniffling detector which further comprises a fast gas chromatography (see, e.g., Snyder et al., 1993). The combination with GC improves the false alarm rate for the detection of dangerous substances by typically one order of magnitude

10.4 Design details

Figs. 10.11 and 10.12 show a portable IMS and GC/IMS detector with a precon-centrator, respectively. GC in combination with IMS significantly reduces the false alarm rate for the detection of hazardous compounds. Preconcentrators (Fig. 10.13) can increase the sample concentration and the sensitivity of the method by a factor of more than 1000, and can also reduce the false alarm rate.

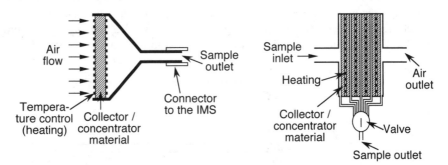

Fig. 10.13   Two design variants of preconcentrators for ion mobility spectrometry. It draws a large volume of air and collects biological and heavy chemical organic compounds from the air onto the filter. The filter is made from zeolites – a material which is commonly used in molecular sieves (see Fig. 10.7). After several minutes of sample collection, the heater vaporizes the organic material into a small parcel of air which is delivered to the IMS. Theses preconcentrators increase the sample concentration by a factor of typically 10–1000 (see, e.g., Spangler, 1992a)

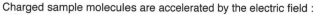

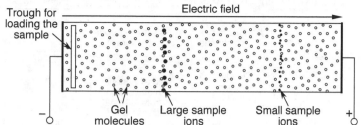

Ions of different sizes and types are differently slowed down by
interaction with gel molecules; typically optical detection

Fig. 10.14 The typical set-up for electrophoresis shows an analogy to ion mobility spectrometry: in both methods charged molecules are accelerated by an electric field and slowed down by interaction with molecules of a stationary phase. In electrophoresis, the stationary phase is generally a gel, and the movement of the sample molecules is often optically detected. A gas serves as stationary phase in ion mobility spectrometry, and the movement of the sample ions is electrically detected. A small size and high charge of ions correlates with a large speed

Ion mobility spectrometry has some similarities to electrophoresis (Fig. 10.14). Because of the similarities to common chromatography, originally ion mobility spectrometry was called "plasma chromatography". However, one has to keep in mind that in contrast to common chromatography, ion mobility spectrometers have a narrow linear range due to space charge effects (Bird and Keller, 1976; Blanchard and Bacon, 1989; Spangler, 1992b), and show serious matrix interferences and prolonged memory effects.

Table 10.1 Common methods of sample ionization in ion mobility spectrometry (Lubman and Kronick, 1982, 1983; Baim et al., 1983; Leasure et al., 1986; Eiceman et al., 1988; Shumate and Hill, 1989; Begley et al., 1991; Phillips and Gormally, 1992; Davies, 1994; Spangler et al., 1994; Carnahan and Tarassov, 1995; Leonhardt, 1996; Lee et al., 1998; Wu et al., 1998a, 1998b, 2000; Budovich et al., 1999; Döring et al., 1999; Borsdorf et al., 2000; Megerle and Cohn, 2000; Schnurpfeil and Klepel, 2000; Borsdorf and Rudolph, 2001)

Method of ionization	Example
Radioactive isotopes	^{3}H, ^{241}Am foil, or ^{63}Ni foil; Fig. 10.15
Photoionization	UV and VUV light from a 30-W krypton or hydrogen lamp with a MgF_2-window, or perpendicular to the drift channel from a frequency-quadrupled Nd:YAG laser at 266 nm (Fig. 10.15). A VUV-absorbing compound may be added for an increased degree of ionization.
Electrospray	Fig. 3.11 in Chap. 3
Laser desorption	Fig. 10.15
Electrical (corona) discharge	Fig. 10.15

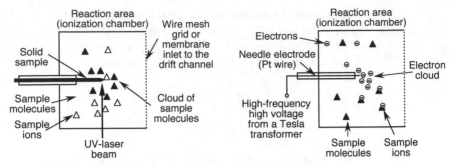

(a) Laser desorption and ionization of a solid sample   (b) Ionization by an electrical discharge

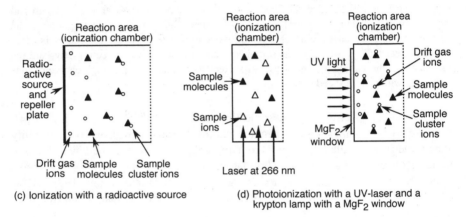

(c) Ionization with a radioactive source

(d) Photoionization with a UV-laser and a krypton lamp with a MgF₂ window

Fig. 10.15 Examples of ionization methods (see also Table 10.1)

Since most of the interesting chemical and biological substances are not charged, it is necessary to ionize them prior to the drift. Table 10.1 and Figs. 10.15 and 10.16 show methods for ionization in ion mobility spectrometry. Radioactive isotopes and photoionization are the most common methods. When using a ^{63}Ni foil as the source of ionization and air as the drift gas, the primary ions are mainly short-living N_2^+, NO^+, and O_2^-. These primary ions rapidly react with traces of water in the drift gas to form clusters of the types $N_2^+(H_2O)_n$, $NO^+(H_2O)_m$, and $O_2^-(H_2O)_k$. Photoionization with hydrogen plasma discharge lamps and krypton plasma discharge lamps requires a photon flux of about 10^{12} cm^{-2} s^{-1}. The geometry of the shutter grid is chosen so that most photons cannot enter the drift channel since this would reduce the resolution. Common radioactive sources in IMS have the advantage of relatively long half-lifes of several years. For immobilization of ^{3}H, it is gettered in a thin titanium layer.

Ion mobility spectrometry has gained significant importance in the context of the detection of ultra-trace chemical and biological contaminants (Snyder et al., 1991a, 1991b, 1996a, 1996b, 1999, 2000; Ogden and Strachan, 1993; Strachan et al., 1995; Dworzanski et al., 1997; Smith et al., 1997), explosives (e.g., Fetterolf and Clark, 1993; Steinfeld and Wormhoudt, 1998; Fig. 10.17), illicit drugs (e.g., Miki et al., 1997, 1998; Keller et al., 1998), pesticides, the detection of animals and animal activity in jungles, and other environmental monitoring.

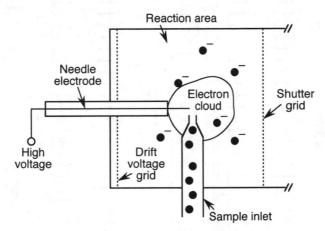

Fig. 10.16 Gas inlet for electrical-discharge (corona discharge) ionization

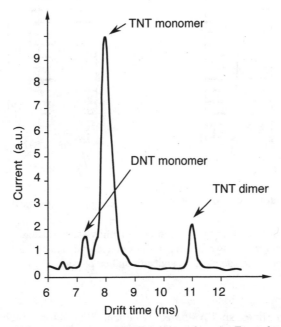

Fig. 10.17 Ion mobility spectrogram of TNT (trinitrotoluene). From data supplied by the Institute for Environmental Technologies Ltd., Berlin

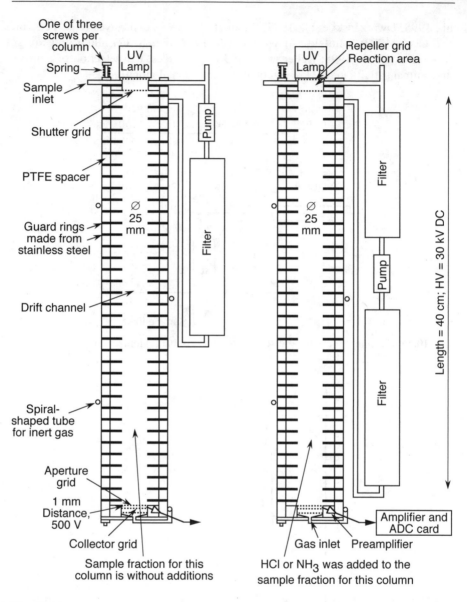

Fig. 10.18 A high resolution two-channel ion mobility spectrometer. Here the sample is simultaneously analyzed in two different ways reducing false identifications of agents. In one of the columns, the sample is chemically and/or physically modified by a chemical addition

Fig. 10.18 outlines an IMS-based twochannel detector which was designed within a feasibility study. In this design the differentiation between various biological and also some chemical substances is improved by utilizing physical and chemical modifications of the sample in one of the two columns, e.g., by adding an acidizing gas: Most proteins display a strong pH-dependency of the charge (see, e.g., Chap. 2 in Nölting, 1999b). Thus, by changing the pH, the charge state of the protein-containing biological material is altered and consequently its speed of diffusion in the drift channel is changed. Further, the added gas can cause chemical changes of some chemical and biological agents. This causes specific changes of the IMS spectra and thus contributes to a further improvement of the correct identification of the agents. Multichannel designs are a further option to reduce false detection rates of IMS. The channels may be operated in the same way speeding up the measurement of slowly drifting substances. Alternatively the sample may simultaneously be distributed over different channels which are operated in various different modes which can improve the resolution of the method.

IMS with an oscillating electric field allow the application of large field strengths without need for a very high voltage (Fig. 10.19). Since very large macromolecules have low mobilities, their fast detection with high resolution requires a high electric field strength in the drift channel. In the common design this may cause safety problems and increases the price of the IMS.

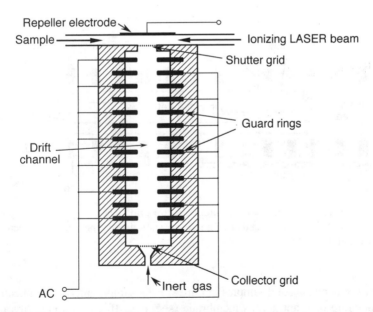

Fig. 10.19 IMS with an oscillating electric field of the guard rings instead of a constant electrostatic field. Only ions of which the movement is in phase with the oscillating electric field can pass through the drift channel and reach the collector. For further details on frequency-domain IMS see, e.g., Martin et al., 1998

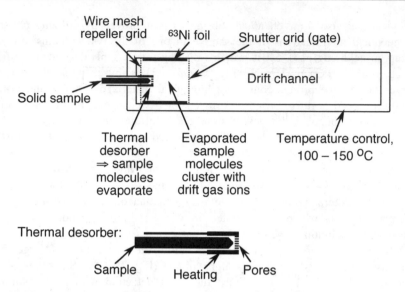

Fig. 10.20 Modified IMS for the detection of chemicals contained in solid samples and for the identification of different solid materials, e.g., wood (Lawrence et al., 1991; Matz and Schröder, 1997; Schröder et al., 1998)

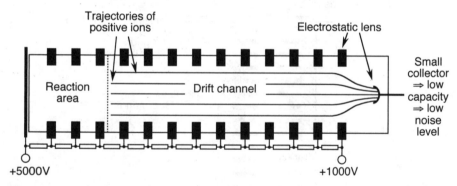

Fig. 10.21 Noise reduction with a smaller collector grid: The collector and the guard ring in front of the collector act together as an electrostatic lens. The smaller collector causes less noise and, thus, a higher sensitivity

Measurements on solid samples require special sample inlets with a heating and a temperature-resistant porous membrane (see, e.g., 10.20) and possibly a higher temperature of the IMS.

An electrostatic lens focusing the ions towards the collector can allow the reduction of the collector size and capacity (Fig. 10.21). This decreases noise and can improve the sensitivity.

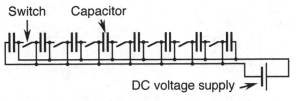

(a) Charging the capacitors

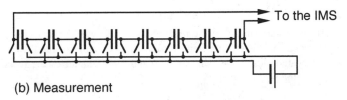

(b) Measurement

Fig. 10.22 Generating the high voltage for an IMS (Goebel and Breit, 2000). (**a**) A set of n capacitors is charged with the voltage V. (**b**) The capacitors are connected together in series for generating the high voltage $n \cdot V$

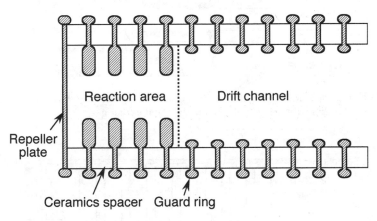

Fig. 10.23 Improved mechanical stability of the drift channel. For a similar design see also (Karl, 1994)

A technique of high voltage generation which does not necessarily require much weight is illustrated in Fig. 10.22: a set of capacitors is charged in parallel and then connected in series.

Parameters which are important for a high reproducibility of IMS spectra are a constant degree of humidity, constant electric field strength, constant source of ionization, efficient removal of previous samples, mechanical stability. The latter can be improved by a special shape of the guard rings (Fig. 10.23). Figs. 10.24–10.26 display some further important innovations in ion mobility spectrometry.

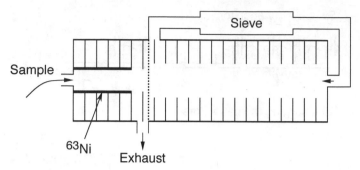

Fig. 10.24 Improved purgeability of the reaction area (see, e.g., Snyder et al., 1993) and improved homogeneity of the electric field the ions experiencing in the drift channel: the sample outlet is located close to the shutter grid, and the diameter of the reaction area is smaller than that of the drift channel

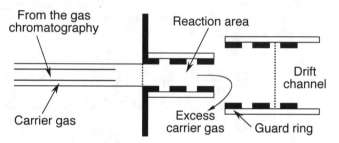

Fig. 10.25 Injection of the output of a gas chromatograph or of the vapor from a solid sample into the IMS with the help of a gentle stream of carrier gas

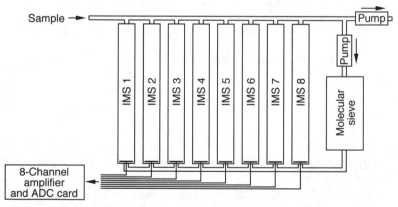

Fig. 10.26 A multichannel ion mobility spectrometer. For an 8 times higher sampling rate than a single-channel spectrometer, all channels are operated in the same way. For decreased rate of false identifications, one sample may be distributed over different channels that are operated in different modes (see, e.g., Turner, 1993)

10.5 Detection of biological agents

Unfortunately, many biological agents are too large to be detected directly: The velocity, v, of a large spherical particle depends on its charge, z, its radius, r, the

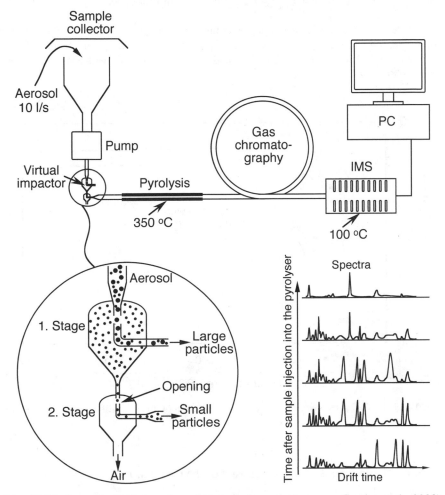

Fig. 10.27 Setup for IMS-detection of biological agents (see, e.g., Snyder et al., 2000). The virtual impactor selects a certain size range of particles, e.g., 1–10 μm, and transfers the selected particles into the pyrolysis tube. Within a few seconds, particles of biological origin are partially decomposed in the pyrolysis reaction at, e.g., 350 °C. In the subsequent analysis of the pyrolysis reaction products, a short gas chromatography and an IMS are combined for enhanced resolution. The operating temperature of the GC/IMS is typically 80–150 °C. For the principle of operation of the virtual impactor see also Fig. 3.20. Due to the highly dispersed 2D-spectra, Py-GC/IMS can potentially much safer unambiguously identify traces of biological agents than a measurement of particle size distribution alone

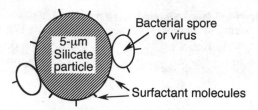

Fig. 10.28 Example of a particle of a dangerous biological contaminant. Light and fluffy composites of bacterial spores or viruses to dust-forming particles of about 1–5 μm diameter can drift in dry air for 100 miles, and can be sucked into the deepest sacs of the lung (Preston, 1998)

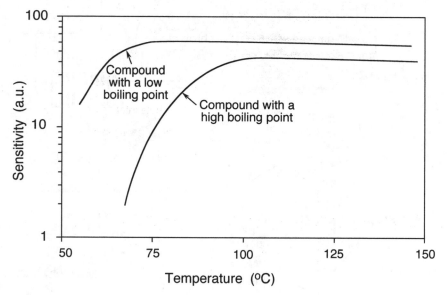

Fig. 10.29 Typical sensitivity of IMS detection of compounds with different boiling points and vapor pressures

electric field strength, E, and the viscosity of the medium, η:

$$v = zE\,(6\pi\eta r)^{-1} \qquad (9.2)$$

Small chemical compounds typically travel in an IMS with several m s^{-1}, but a single-charged dust particle ($z = 1.6 \times 10^{-19}$C) with a radius of 1 μm travels in air ($\eta_{air} = 1.8 \times 10^{-5}$ N s m^{-2}) at a field strength of $E = 300$ V mm^{-1} only with about 0.14 mm s^{-1}. At higher gating frequencies of the IMS, successive spectra of slowly moving agents would superimpose. Also, just average size and size distribution of biological agents offers little information about the precise nature of the agent. That is why biological agents are pyrolyzed prior to analysis in the IMS (Fig. 10.27). Pyrolysis (see also Sect. 3.2) decomposes and vaporizes biological agents and can be applied on bacteria and viruses (Fig. 10.28). The low vapor pressure of most biological compounds requires an operation of the IMS at a sufficiently high temperature (Fig. 10.29). The set-up virtual impactor / pyrolyzer / GC / IMS (Fig. 10.27) is capable to detect a few bacterial spores in a volume of several 1000 liters.

11 Microwave auditory effects and the theoretical concept of thought transmission technology

It should be pointed out that intense electromagnetic radiation can be hazardous and should be handled with great care and always according to all safety regulations and laws. The currently ongoing discussion on electrosmog underlines the importance to use electromagnetic radiation in a responsible manner. These are only some reasons why every application of thought transmission technology always has to be in strict accordance with all legal requirements and ethical guidelines. Of course, this applies also to all other scientific methods.

11.1 Theoretical concept

The present theoretical concept of thought transmission technology was designed to overcome the limitations of traditional methods for broadcasting information, such as radio, television, and mobile telephones, e.g., to some carefully selected persons in cases of emergencies, and is partially based on the perceptibility of microwaves (Frey, 1962, 1993; Lin, 1978, 2000; Nölting, 2002). The concept shows the theoretical feasibility of thought transmissions. It discloses methods and apparatuses for thought transmission via electromagnetic waves whereby the receiving person does not need any electronic devices for the reception. According to the concept, thought transmission is performed by sending focussed low-energy modulated electromagnetic energy into the organism of the receiving person in such a way that it changes the probability of certain thoughts. In one embodiment of the methods and apparatuses of the concept, (a) electrical signals of voices are stored on a computer, (b) a computer transfers the signals into a sequence of electrical pulses of which the envelope correlates with the intensity course of the voices, (c) a maser beam is modulated with the obtained sequence of pulses, (d) the maser beam is aimed at the receiving person and an intensity is chosen which is just below the intensity which consciously can be perceived. In another embodiment of the methods and apparatuses of the concept, (a) a large number of correlations between acoustical stimuli and reactions or thoughts is measured, (b) a computer uses the set of correlations to generate a sequence of stimuli which would cause a certain reaction of thought with high probability, (c) the sequence is modulated onto a focussed beam of electromagnetic energy, (d) the beam is sent to the receiving person at an subliminal or perceivable intensity and duration, (e) the reactions of the person are monitored with millimeterwave cameras and the intensity adjusted if necessary. The invention of microwave auditory and thought transmission technology could be applied, e.g., (a) to support the communication with persons buried by rubble after earth quakes, (b) for unobtrusive transmissions of information to security personnel, (c) to support important people presenting lectures in public, (d) to sensitize persons regarding important topics in emergency situations, e.g., by sending warnings, and to protect the population form catastrophic events, (e) in combination with detection methods for thought reading on consenting people to contribute to the understanding of the function of the human brain, (f) to support essential learning and re-learning when other methods are not applicable, (g) for the therapy and prophylaxis of certain pathological impairments of the brain metabolism and to influence certain non-pathological limitations and stress situations.

11.1.1 Background of the concept of thought transmission technology

Modern media such as radio and television require an electronic device transforming electromagnetic energy into a perceivable acoustical or optical signal, and quite often persons can not individually be provided with information.

On the other hand since the 1940's it is known that certain RADAR pulses can be perceived as a knocking sound. This phenomenon was named microwave auditory effect. Since the 1960's, a large number of studies has been undertaken regarding the audibility of modulated microwave energy if irradiated into the head (Fig. 11.1; Frey, 1961, 1962, 1993; Frey and Messener, 1973; Lin, 1978, 1989;

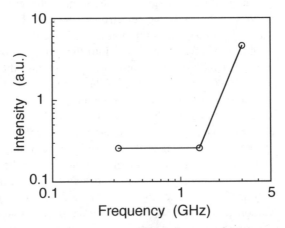

Fig. 11.1 Relative intensity which is necessary under some experimental conditions to perceive pulse modulated microwave energy (Lin, 1978)

Frey and Corin, 1979; Brunkan, 1989; Stocklin, 1989). Also known are methods of affecting emotions with acoustical or electrical stimulation (Meland, 1980; Gall, 1994), and the application of acoustical signals for the subconscious influencing of humans (Lowery, 1992).

The concept of thought transmission represents a further development of these findings and inventions as outlined in the next section.

11.1.2 Description of the technology

According to the theoretical concept, remote thought transmission is based on the use of directional radio. In contrast to common directional radio, the electromagnetic beam is directly coupled into the organism of the receiving person, e.g., the head, big cerebral cortex, inner ear, auditory nerves, or other nerves. Depending on the type of signal modulated onto the electromagnetic beam, this coupling causes an intended change of the thoughts of the receiving person. In general the change of thoughts is only statistically effective, e.g., the probability for certain thoughts is increased or decreased in a predictable way. However, in some cases the change may be deterministic. Thought transmission is suitable in some cases for combination with observations by millimeterwave cameras and microwave based voice transmission in which the conscious audibility of modulated microwave energy is used, but may also be used independently of these methods.

For example, in a simple type of a thought transmission apparatus, (a) the operator speaks words into a microphone, (b) the electrical signals of the microphone are transformed into sequences of pulses, e.g., of 100 μs duration and

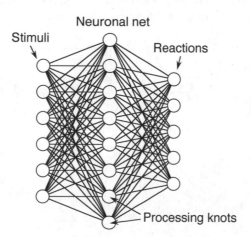

Fig. 11.2 Example of a neuronal net integrated into the software of a thought transmission apparatus. It is used to calculate the sets of stimuli necessary to cause certain reactions or thoughts. The neuronal knots were initially trained with a large number of reactions to certain stimuli

200 μs separation, (c) the sequences of pulses are stored on a computer, (d) sequences of pulses which correspond to the thoughts to be sent are retrieved from the computer, (e) a microwave beam is modulated with the retrieved sequence of pulses and sent to the receiving person, whereby the intensity is so low or the used frequencies or signal durations are such that the receiving person has no conscious perception of the sending, (f) the subconscious action of the microwave beam is observed by the operator.

In a more complicated design of a thought transmission apparatus, for example, the operator enters the thought to be sent into a computer which translates, with the help of tables or neural networks (Fig. 11.2), the thought into a sequence of signals that are modulated onto the microwave beam which is sent to the receiving person. This sequence of signals may contain microwave induced consciously and subconsciously perceivable sounds. The calculation of a table for the translation between thoughts and sequences of signals inducing these thoughts may be done with the help of a set of correlations between stimuli and reactions. Likewise the training of neural networks for the translation of thoughts into sequences of signals is carried out under observation of reactions to a set of stimuli. A sophisticated computer-aided thought transmission apparatus should involve automatic or semi-automatic operation of thought transmission and beam tracking.

11.1.2.1 Frequencies

Carrier frequencies around 100 MHz – 100 GHz appear to be particularly suitable: electromagnetic waves of this frequency range travel linearly, can be focussed, penetrate air and walls of buildings, and induce currents in the outer layers of the human body (Fig. 11.3). For a good efficiency of thought transmission, the carrier frequency may match the resonance frequency of body parts, e.g., certain nerves.

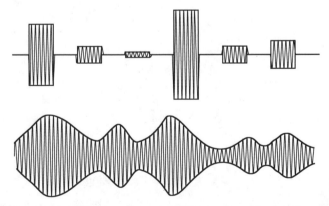

Fig. 11.3 Two examples for amplitude modulated signals. The envelope of the signal corresponds to a utilizable low frequency signal which is effective, e.g., in the cerebral cortex, inner ear, auditory nerves, or other nerves and typically has a frequency of 1 Hz – 100 MHz

Theoretical studies and measurements of the audibility of microwaves suggest that frequencies in the range of 1 Hz – 100 MHz for the signals modulated onto the carrier frequency should to be particularly suitable. For example, voice signals in the audible range of 16 Hz – 20 kHz may contain components in the MHz region when transformed into pulse sequences. Low frequencies, e.g., in the range of 1–60 Hz are known to be effective in mood influencing.

11.1.2.2 Radiation sources

For the required beam properties and frequency ranges, masers (microwave amplification by stimulated emission of radiation; Fig. 11.4), e.g., free electron masers (often also called free electron lasers) and semiconductor diode masers (diode lasers), are suitable sources. Also magnetrons, gyrotrons, klystrons and phased arrays are suitable for the generation of a focussed, modulated electromagnetic beam with a frequency suitable for transmission over several meters to kilometers.

Depending on the magnitude of the transmission losses and the body part of interaction, the necessary average power per addressed person is predicted to be in most cases of the order of 0.01–10 W. Larger intensities might be necessary for transmissions through walls, but great care has to be taken to avoid injuries to the receiving or other persons in the beam path.

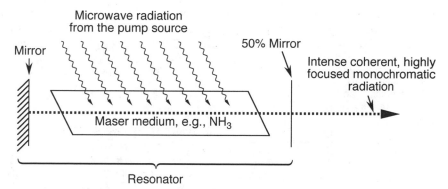

Fig. 11.4 Maser (microwave-amplified stimulated emission of radiation) principle of generating and amplifying electromagnetic waves. The pump source excites specific lines in the maser medium. Under certain conditions, the energy level populations in the masing molecules of the maser medium become inverted, i.e. there are more molecules in an excited metastable state than in the ground state. Under these conditions of population inversion, microwaves traveling through the maser medium can become enormously amplified in intensity by induced emission. If the amplification and quality of the resonator are sufficient, starting from spontaneous emission events, self-sustained oscillations result and no input beam is required

11.2 Examples of potential applications

11.2.1 An automated device

Fig. 11.5 outlines a thought transmission device mounted on a vehicle. It generates a microwave beam which is modulated in a suitable way and sent to the receiving person. The conceived device comprises a maser for the generation of the electromagnetic beam, a microphone for the input of voice signals by the operator, a rechargeable source of energy, and a detector for observation and tracking, and has a total weight of about 100 kg. In order to ensure both a good

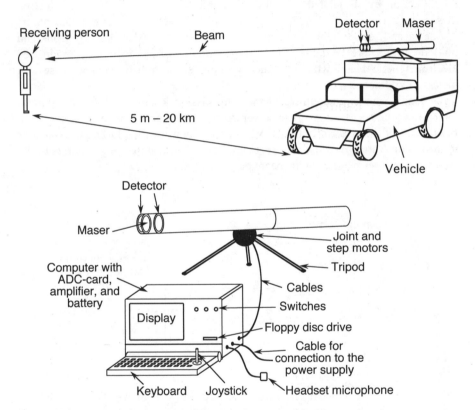

Fig. 11.5 Remote thought transmission to a receiving person by a modulated beam of microwaves of millimeterwaves generated by a maser, e.g., a semiconductor maser, mounted on top of a recognizance vehicle. In one mode, the voice of the observer is changed into an electrical signal and feeded into the computer which further processes the voice signals. The maser beam is modulated with these processed signals and sent to the receiving person. The modulation may comprise, e.g., subliminal sounds, consciously perceivable sounds, and low frequency mood changing signals. The tracking of the receiving person may be carried out manually or automatically. Millimeterwave telescopes or RADAR detectors should be suitable for the detection of the receiving person

focusability and sufficient penetration through air, walls of buildings and soil, the carrier frequency of the maser should be in the region of 1–1000 GHz. Ideally the detector is sensitive at the carrier frequency of the maser.

The thought transmission is carried out, e.g., by sending a sequence of microwave pulses of which the intensities correlate with the sound intensity of certain keywords. The microwave beam induces currents in the cerebral cortex of the receiving person which change the probability of certain thoughts. The whole apparatus may operate automatically by using a software for tracking and sending the desired signals. For the development of the software which generates the modulation signals, one may use a large number of weak correlations between stimuli and thoughts. For example, if 100 independent stimuli cause a 2%-probability of a certain thought, combined these stimuli may cause a 87%-probability of that thought. Since in some applications thought transmission may be applied over a relatively long period of time, it may be possible to use such weak correlations to obtain a significant result.

11.2.2 A small handheld device

Fig. 11.6 outlines a handheld thought transmission device comprised of a maser, headset with microphone for the input of voice signals by the operator (observer), a rechargeable battery, and a detector, e.g., a millimeterwave camera. The opera-

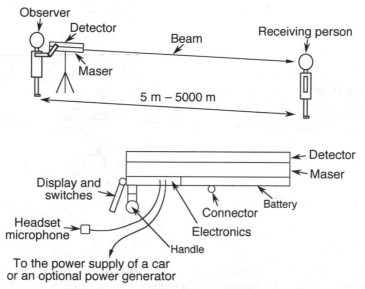

Fig. 11.6 An example of a small handheld thought transmission device comprised of a maser, a headset with microphone, a rechargeable battery, and a detector. The thought transmission device can be connected with the power supply of a car or with a power generator. By means of a display and handle, the receiving person is tracked

tor tracks the receiving person manually and sends stored sequences of signals or just the electronically processed voice modulated onto the maser beam. Several switches and the electronics enable the operator the setting of various modes, such as the sending of stored signals, semi-automatic adjustment of intensity, type of modulation for the transmission of sounds. In contrast to stunning shots with electromagnetic guns, the intensity of radiation is much lower and may be below or just above the level of conscious perception. Besides perceivable and subconscious voice, music, and rhythms in the frequency range above 16 Hz, also low frequency rhythms below 16 Hz may be sent.

With the help of this device, important presentations might be supported: for example, a team of experts watches the presentation of an important person and supports it in important situations by sending keywords or other advice.

11.2.3 Biomedical microwave auditory and thought transmissions

Figs. 11.7–11.15 illustrate several design variants and the possibility of placing microwave auditory and thought transmission technology at various locations to reach different sites on earth. To some degree, microwave auditory and thought transmission technology should be linkable with existing networks of telecommunication technology (Fig. 11.13).

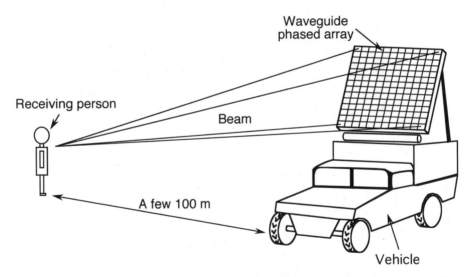

Fig. 11.7   Phased array antenna thought transmission device mounted on top of a recognizance vehicle. The phased array antenna tracks the receiving person without mechanical movement of the antenna, similarly to a phased array RADAR. For thought transmission, e.g., sequences of pulses with intensities that correlate with the intensity of spoken words are send with low intensities

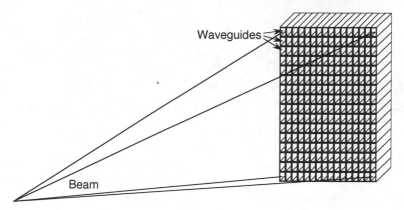

Fig. 11.8 A small phased array antenna with 16×16 waveguides. The elements of the array are made using, e.g., semiconductor diodes. The direction of the beam is changed by changing the phases of the elements of the array

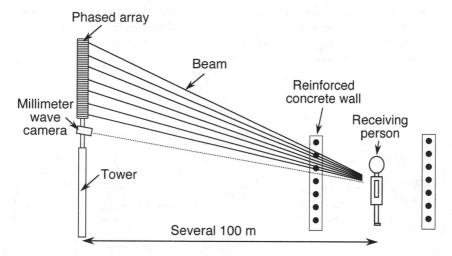

Fig. 11.9 Phased arrays allow thought transmission and observation through enforced concrete walls: because of the conical geometry of the beam, small pieces of metal in the beam path do not present a significant problem

In case of an emergency or a catastrophe, a considerable number of people could be contacted without delay via microwave auditory and thought transmission (Fig. 11.14). In such situations, the transmission could support damage limitation, panic prevention and could help to motivate people to a support rescue operations. The requirement of fast beam direction changes can be met, e.g., by masers with electronic direction modulators and phased arrays.

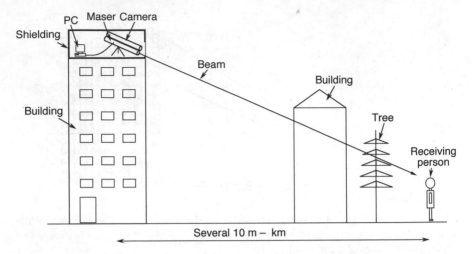

Fig. 11.10 Remote thought transmission via a modulated maser beam over very large distances. The camera may detect, e.g., the reflected maser radiation. The range depends on the beam divergence and relief of the landscape. A good shielding of the room which contains the apparatus may be necessary to prevent unintended interference with humans and electrical equipment. To some degree the transmission and observation can penetrate trees and walls from wood, stone, and plastics

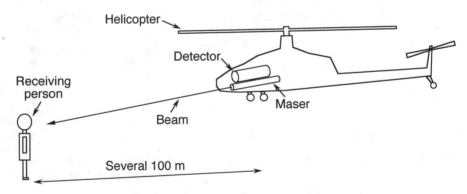

Fig. 11.11 Thought transmission from a plane, drone, or helicopter by a specifically modulated beam of electromagnetic energy. The transmission may require special movement stabilizers of the thought transmission apparatus

Person-to-person thought transmission is conceivable: the signal to be sent is taped from the head of one person, processed on a computer, and sent to another person. The processing may involve a frequency analysis and selection of the predominant frequencies. The transmission may involve non-audible very low or very high frequencies. In this way, states of tension or relaxation which differ by their predominant frequencies of brain activity may be transmitted. If the sending person is, e.g., a patient in coma, the receiving healthy person can establish a new form of contact with the patient. The method might also help to improve contact with blind deaf and dumb patients.

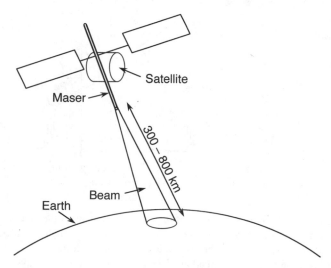

Fig. 11.12 Thought transmission via satellite requires a very low beam divergence and strong power supply. Satellites in low orbits could transmit for a couple of minutes to a certain spot on earth before being out of range. Problematical would be, however, the requirements for a sufficient power supply and moderate weight of the transmitter

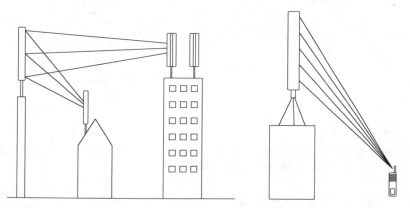

Fig. 11.13 To some degree the thought transmission technology should be linkable with the existing network of phased array antennas that allow mobile phone applications throughout the whole industrialized world

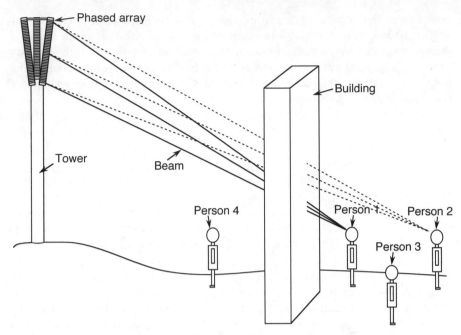

Fig. 11.14 Thought transmission to several important persons with the help of specifically modulated electromagnetic beams in case of an emergency. For better detection, the receiving people carry labels. The transmission is carried out multiplexed, i.e. almost simultaneously by fast switching of the direction of the beams of the phased arrays

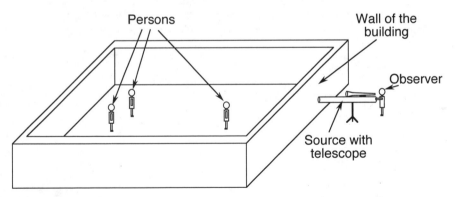

Fig. 11.15 Microwave auditory and thought transmission in case of an emergency with the help of an electromagnetic source operated at low intensity

Another conceivable application of thought transmission is profiling and thought reading on consenting people wrongly suspected of a crime wanting to prove their innocence: A simple method would be to surprisingly and subconsciously sending the person a keyword which has an important meaning in the context of a crime. By simultaneous observation, a suspicion is substantiated or unsubstantiated. A preparation phase or sensitization phase in which the thoughts of the person are directed to the key event by subliminal signals may precede the transmission of the keyword. Computer-aided thought transmission technology should allow even more sophisticated methods: for example, certain key information might subconsciously be sent over certain periods of time with varying intensities and the correlation of the reactions of the receiving person with the signal be calculated.

Microwave auditory and thought transmission technology might also be expected to play an increasing role in brain research, in the treatment of diseases, and for the analysis of biochemical networks in the brain. For example, since the modulated electromagnetic radiation of thought transmission may effect the brain and certain neurological processes in a different way than sound waves or visible light, and also non-audible frequencies may be used, new medical applications are feasible. So the methods of microwave auditory transmission and thought transmission should enable new ways of analysis, therapy and prophylaxis of certain pathological limitations of brain metabolism and be useful to influence certain non-pathological limitations, stress situations, and possibly losses of function of the brain. In this way microwave auditory transmission and thought transmission technology might not only provide an increasingly important tool for communication, but also for understanding the human brain.

12 Conclusions

Biophysical methods are extremely important for the further understanding of biological processes. Structurally highly resolving methods such as X-ray crystallography and scanning probe microscopy in combination with kinetic methods (see, e.g., Nölting, 1999b) give us a true understanding of biological processes at a molecular and cellular level. Proteome maps of healthy and ill individuals are compared for identification of up- or down-regulation in diseases states and for individual, highly efficient drug targeting. Biophysical nanotechnology and mass spectrometry open new fascinating ways of studying and influencing complex biological systems. Biophysical nanotechnology takes novel approaches to assemble protein nanoarrays, nanoparticles, and nanowires to well functioning structures. Mass spectrometry and ion mobility spectrometry have significantly advanced towards the detection of ultra-traces not only of chemical, but also of biological agents.

The biophysical understanding of the living world is crucial to develop rational strategies to influence, in a responsible manner, pathological and non-pathological limitations, stress situations, and diseases states. In this way, biophysics can contribute to the understanding of the factors that affect the prosperity and evolution of the human society.

References

Abkevich VI, Gutin AM, Shakhnovich EI (1995) Impact of local and nonlocal interactions on the thermodynamics and kinetics of protein folding. J Mol Biol 252:460–471

Abola EE, Sussman JL, Prilusky J, Manning NO (1997) Protein data bank archives of three-dimensional macromolecular structures. Methods Enzymol 277:556–571

Achari A, Hale SP, Howard AJ, Clore GM, Gronenborn AM, Hardman KD, Whitlow M (1992) 1.67 Å X-ray structure of the B2 immunoglobulin-binding domain of streptococcal protein G and comparison to the NMR structure of the B1 domain. Biochemistry 31:10449–10457

Adam GC, Sorensen EJ, Cravatt BF (2002a) Chemical strategies for functional proteomics. Mol Cell Proteomics 1:781–790

Adam GC, Sorensen EJ, Cravatt BF (2002b) Proteomic profiling of mechanistically distinct enzyme classes using a common chemotype. Nat Biotechnol 20:805–809

Aitio H, Laakso T, Pihlajamaa T, Torkkeli M, Kilpelainen I, Drakenberg T, Serimaa R, Annila A (2001) Characterization of apo and partially saturated states of calerythrin, an EF-hand protein from *S. erythraea*: a molten globule when deprived of Ca^{2+}. Protein Sci 10:74–82

Aitken A, Learmonth M (2002) Protein identification by in-gel digestion and mass spectrometric analysis. Mol Biotechnol 20:95–97

Allison DP, Hinterdorfer P, Han W (2002) Biomolecular force measurements and the atomic force microscope. Curr Opin Biotechnol 13:47–51

Alm E, Baker D (1999) Matching theory and experiment in protein folding. Curr Opin Struct Biol 9:189–196

Andersen KV, Poulsen FM (1992) Three-dimensional structure in solution of acyl-coenzyme A binding protein from bovine liver. J Mol Biol 226:1131–1141

Anderson RJ, Bendell DJ, Garnett I, Groundwater PW, Lough WJ, Mills MJ, Savery D, Shattock PE (2002) Identification of indolyl-3-acryloylglycine in the urine of people with autism. J Pharm Pharmacol 54:295–298

Andolfi L, Cannistraro S, Canters GW, Facci P, Ficca AG, van Amsterdam IM, Verbeet MP (2002) A poplar plastocyanin mutant suitable for adsorption onto gold surface via disulfide bridge. Arch Biochem Biophys 399:81–88

Andretzky P, Lindner MW, Herrmann JM, Schultz A, Konzog M, Kiesewetter F, Häusler G (1998) Optical coherence tomography by "Spectral radar". SPIE 3567:78–87

Åqvist J (1999) Long-range electrostatic effects on peptide folding. FEBS Lett 457:414–418

Arai S, Hirai M (1999) Reversibility and hierarchy of thermal transition of hen egg-white lysozyme studied by small angle X-ray scattering. Biophys J 76:2192–2197

Aries RE, Gutteridge CS, Ottley TW (1986) Evaluation of a low-cost, automated pyrolysis mass spectrometer. J Anal Appl Pyrol 9:81–98

Armitage S, Saywell S, Roux C, Lennard C, Greenwood P (2001) The analysis of forensic samples using laser micro-pyrolysis gas chromatography mass spectrometry. J Forensic Sci 46:1043–1052

Asbury GR, Hill HH (2000) Separation of amino acids by ion mobility spectrometry. J Chromatogr A 902:433–437

Ash EA, Nicholls G (1972) Super-resolution aperture scanning microscope. Nature 237:510–513

Astley OM, Donald AM (2001) A small angle X-ray scattering study of the effect of hydration on the microstructure of flax fibers. Biomacromolecules 2:672–680

Avseenko NV, Morozova TY, Ataullakhanov FI, Morozov VN (2001) Immobilization of proteins in immunochemical microarrays fabricated by electrospray deposition. Anal Chem 73:6047–6052

Avseenko NV, Morozova TY, Ataullakhanov FI, Morozov VN (2002) Immunoassay with multicomponent protein microarrays fabricated by electrospray deposition. Anal Chem 74:927–933

Backmann J, Schäfer G, Wyns L, Bönisch H (1998) Thermodynamics and kinetics of unfolding of the thermostable trimeric adenylate kinase from the archaeon *Sulfolobus acidocaldarius*. J Mol Biol 284:817–833

Bada M, Walther D, Arcangioli B, Doniach S, Delarue M (2000) Solution structural studies and low-resolution model of the *Schizosaccharomyces pombe* sap1 protein. J Mol Biol 300:563–574

Bai YW (1999) Equilibrium amide hydrogen exchange and protein folding kinetics. J Biomol NMR 15:65–70

Bai YW (2000) Kinetic evidence of an on-pathway intermediate in the folding of lysozyme. Protein Sci 9:194–196

Bailey JA, Tomson FL, Mecklenburg SL, MacDonald GM, Katsonouri A, Puustinen A, Gennis RB, Woodruff WH, Dyer RB (2002) Time-resolved step-scan Fourier transform infrared spectroscopy of the CO adducts of bovine cytochrome c oxidase and of cytochrome bo_3 from *Escherichia coli*. Biochemistry 41:2675–2683

Baim MA, Eatherton RL, Hill HH (1983) Ion mobility detector for gas chromatography with a direct photoionization source. Anal Chem 55:1761–1766

Baker D (2000) A surprising simplicity to protein folding. Nature 405:39–42

Baker D, DeGrado WF (1999) Engineering and design – editorial overview. Curr Opin Struct Biol 9:485–486

Barrera FN, Garzon MT, Gomez J, Neira JL (2002) Equilibrium unfolding of the C-terminal SAM domain of p73. Biochemistry 41:5743–5753

Barshick SA, Wolf DA, Vass AA (1999) Differentiation of microorganisms based on pyrolysis ion trap mass spectrometry using chemical ionization. Anal Chem 71:633–641

Barth A (2002) Selective monitoring of 3 out of 50,000 protein vibrations. Biopolymers 67:237–241

Baumbach JI, Eiceman GA (1999) Ion mobility spectrometry: arriving on site and moving beyond a low profile. Appl Spectrosc 53:338A–355A

Baumbach JI, Stach J (ed) (1998) Recent Developments in Ion Mobility Spectrometry. International Society for Ion Mobility Spectrometry, Dortmund, Germany

Beamer LJ, Pabo CO (1992) Refined 1.8 Å crystal structure of the λ-repressor-DNA complex. J Mol Biol 227:177–196

Beegle LW, Kanik I, Matz L, Hill HH (2001) Electrospray ionization high-resolution ion mobility spectrometry for the detection of organic compounds, 1. Amino acids. Anal Chem 73:3028–3034

Begley P, Corbin R, Foulger BE, Simmonds PG (1991) Photoemissive ionization source for ion mobility detectors. J Chromatogr 588:239–249

Bellotti V, Stoppini M, Mangione P, Sunde M, Robinson C, Asti L, Brancaccio D, Ferri G (1998) β2-microglobulin can be refolded into a native state from *ex vivo* amyloid fibrills. Eur J Biochem 258:61–67

Benoit M, Gabriel D, Gerisch G, Gaub HE (2000) Discrete interactions in cell adhesion measured by single-molecule force spectroscopy. Nature Cell Biol 2:313–317

Berkeley RCW, Goodacre R, Helyer RJ, Kelley T (1990) Pyrolysis-MS in the identification of micro-organisms. Lab Pract 39:81–83

Berks BC, Sargent F, Palmer T (2000) The Tat protein export pathway. Mol Microbiol 35:260–274

Bernal JD (1939) The structure of proteins. Nature 143:663–667

Bernal JD, Crowfoot D (1934) X-ray photographs of crystalline pepsin. Nature 138:133–134

Betzig E, Harootunian A, Isaacson M, Kratschmer E, Lewis A (1986) Near-field scanning optical microscopy (NSOM): development and biophysical applications. Biophys J 49:269–279

Betzig E, Trautmann JK, Harris TD, Weiner JS, Kostelak RL (1991) Breaking the diffraction barrier: optical microscopy on a nanometric scale. Science 251:1468–1470

Betzig E, Finn PL, Weiner JS (1992) Combined shear force and near-field scanning optical microscopy. Appl Phys Lett 60:2484–2486

Bianco A, Corvaja C, Crisma M, Guldi DM, Maggini M, Sartori E, Toniolo C (2002) A helical peptide receptor for [60]fullerene. Chemistry 8:1544–1553

Binnig G, Rohrer H (1987) Scanning tunneling microscopy – from birth to adolescence. Rev Mod Phys 59, No 3, Part I

Binnig G, Gerber C, Quate CF (1986) Atomic force microscopy. Phys Rev Lett 56:930–933

Binnig G, Rohrer H, Gerber C, Weibel E (1982a) Vacuum tunneling. Physica 109 & 110B 2075–2077

Binnig G, Rohrer H, Gerber C, Weibel E (1982b) Surface studies by scanning tunneling microscopy. Phys Rev Lett 49:57–61

Binnig G, Rohrer H, Gerber C, Weibel E (1983) 7×7 reconstruction on Si(111) resolved in real space. Phys Rev Lett 50:120–123

Bird GM, Keller RA (1976) Vapor concentration dependence of plasmagrams. J Chromatographic Sci 14:574–577

Birolo L, Dal Piaz F, Pucci P, Marino G (2002) Structural characterization of the M* partly folded intermediate of wild-type and P138A aspartate aminotransferase from *Escherichia coli*. J Biol Chem 277:17428–17437

Biscarini F, Cavallini M, Leigh DA, Leon S, Teat SJ, Wong JK, Zerbetto F (2002) The effect of mechanical interlocking on crystal packing: predictions and testing. J Am Chem Soc 124:225–233

Blanchard WC, Bacon AT (1989) Ion mobility spectrometer. US Patent 4,797,554

Blomberg A (2002) Use of two-dimensional gels in yeast proteomics. Methods Enzymol 350:559–584

Boettner M, Prinz B, Holz C, Stahl U, Lang C (2002) High-throughput screening for expression of heterologous proteins in the yeast *Pichia pastoris*. J Biotechnol 99:51–62

Boisvert DC, Wang J, Otwinowski Z, Horwich AL, Sigler PB (1996) The 2.4 Å crystal Structure of the bacterial chaperonin GroEL complexed with ATP γ S Nature Struct Biol 3:170–177

Bork P (2002) Comparative analysis of protein interaction networks. Bioinformatics 18 Suppl 2:S64

Borsdorf H, Rudolph M (2001) Gas-phase ion mobility studies of constitutional isomeric hydrocarbons using different ionization techniques. Int J Mass Spectrom 208:67–72

Borsdorf H, Schelhorn H, Flachowsky J, Döring HR, Stach J (2000) Corona discharge ion mobility spectrometry of aliphatic and aromatic hydrocarbons. Anal Chim Acta 403:235–242

Braig K, Otwinowski Z, Hegde R, Boisvert DC, Joachimiak A, Horwich AL, Sigler PB (1994) The crystal structure of the bacterial chaperonin GroEL at 2.8 Å. Nature 371:578–586

Brask J, Jensen KJ (2000) Carbopeptides: chemoselective ligation of peptide aldehydes to an aminooxy-functionalized D-galactose template. J Pept Sci 6:290–299

Brown DR, Wong BS, Haifiz F, Clive C, Haswell SJ, Jones IM (1999) Normal prion protein has an activity like that of superoxide dismutase. Biochem J 344:1–5

Brown DR, Hafiz F, Glasssmith LL, Wong BS, Jones IM, Clive C, Haswell SJ (2000) Consequences of manganese replacement of copper for prion protein function and proteinase resistance. EMBO J 19:1180–1186

Bruckbauer A, Ying L, Rothery AM, Korchev YE, Klenerman D (2002a) Characterization of a novel light source for simultaneous optical and scanning ion conductance microscopy. Anal Chem 74:2612–2616

Bruckbauer A, Ying L, Rothery AM, Zhou D, Shevchuk AI, Abell C, Korchev YE, Klenerman D (2002b) Writing with DNA and protein using a nanopipet for controlled delivery. J Am Chem Soc 124:8810–8811

Brunkan WB (1989) Hearing system. US Patent 4,877,027

Bryden WA (1995) Tiny-TOF-MALDI mass spectrometry for particulate drug and explosives detection. 3rd Symposium on Research & Development, Kossiakoff Center, The Johns Hopkins University

Budovich VL, Mikhailov AA, Arnold G (1999) Ion mobility spectrometer. US Patent 5,969,349

Burbaum J, Tobal GM (2002) Proteomics in drug discovery. Curr Opin Chem Biol 6:427–433

Burke JR (1992) Ion mobility detector. US Patent 5,162,649

Burton RE, Huang GS, Daugherty MA, Fullbright PW, Oas TG (1996) Microsecond protein folding through a compact transition state. J Mol Biol 263:311–322

Burton RE, Huang GS, Daugherty MA, Calderone TL, Oas TG (1997) The energy landscape of a fast-folding protein mapped by Ala→Gly substitutions. Nature Struct Biol 4:305–310

Bustamante C, Smith SB, Liphardt J, Smith D (2000) Single-molecule studies of DNA mechanics. Curr Opin Struct Biol 10:279–285

Butler BC, Hanchett RH, Rafailov H, MacDonald G (2002) Investigating structural changes induced by nucleotide binding to RecA using difference FTIR. Biophys J 82:2198–2210

Calvo EJ, Danilowicz C, Wolosiuk A (2002) Molecular "wiring" enzymes in organized nanostructures. J Am Chem Soc 124:2452–2453

Campbell DN, Spangler GE, Davis RC Jr, Fafaul EF, Carrico JP Jr (1991) All ceramic ion mobility spectrometer cell. US Patent 5,021,654

Canady MA, Tsuruta H, Johnson JE (2001) Analysis of rapid, large-scale protein quaternary structural changes: time-resolved X-ray solution scattering of *Nudaurelia capensis* Ω virus (NΩV) maturation. J Mol Biol 311:803–814

Canet D, Last AM, Tito P, Sunde M, Spencer A, Archer DB, Redfield C, Robinson CV, Dobson CM (2002) Local cooperativity in the unfolding of an amyloidogenic variant of human lysozyme. Nature Struct Biol 9:308–315

Carnahan BL, Tarassov AS (1995) Ion mobility spectrometer. US Patent 5,420,424

Carnahan BL, Tarassov AS (1998) Recirculating filtration system for use with a transportable ion mobility spectrometer. US Patent 5,723,861

Caroll DI (1972) Apparatus and methods for separating, detecting, and measuring trace gases. US Patent 3,668,383

Caroll DI, Cohen MJ, Wernlund RF (1971) Apparatus and methods for separating, detecting, and measuring trace gases with enhanced resolution. US Patent 3,626,180

Carrion-Vazquez M, Oberhauser AF, Fowler SB, Marszalek PE, Broedel SE, Clarke J, Fernandez JM (1999) Mechanical and chemical unfolding of a single protein: a comparison. Proc Natl Acad Sci USA 96:3694–3699

Casari G, Sippl MJ (1992) Structure-derived hydrophobic potential – hydrophobic potential derived from X-ray structures of globular proteins is able to identify native folds. J Mol Biol 224:725–732

Castellanos IJ, Cruz G, Crespo R, Griebenow K (2002) Encapsulation-induced aggregation and loss in activity of γ-chymotrypsin and their prevention. J Control Release 81:307–319

Celis JE, Celis P, Palsdottir H, Ostergaard M, Gromov P, Primdahl H, Orntoft TF, Wolf H, Celis A, Gromova I (2002) Proteomic strategies to reveal tumor heterogeneity among urothelial papillomas. Mol Cell Proteomics 1:269–279

Chacon P, Moran F, Diaz JF, Pantos E, Andreu JM (1998) Low-resolution structures of proteins in solution retrieved from X-ray scattering with a genetic algorithm. Biophys J 74:2760–2775

Chambert R, Petit-Glatron MF (1999) Anionic polymers of *Bacillus subtilis* cell wall modulate the folding rate of secreted proteins. FEMS Microbiol. Lett 179:43–47

Chan HS (1998) Protein folding: matching speed and locality. Nature 392:761–763

Chan HS (1999) Folding alphabets. Nature Struct Biol 6:994–996

Chan HS (2000) Modeling protein density of states: additive hydrophobic effects are insufficient for calorimetric two-state cooperativity. Proteins 40:543–571

Chang AM, Hallen HD, Harriott L, Hess HF, Kao HL, Kwo J, Miller RE, Wolfe R, van der Ziel J, Chang TY (1992a) Scanning Hall probe microscopy. Appl Phys Lett 61:1974–1976

Chang AM, Hallen HD, Hess HF, Kao HL, Kwo J, Sudbø A, Chang TY (1992b) Scanning Hall probe microscopy of a vortex and field fluctuations in $La_{1.85}Sr_{0.15}CO_4$ films. Europhys Lett 20:645–650

Chang JY, Li L, Bulychev A (2000a) The underlying mechanism for the diversity of disulfide folding pathways. J Biol Chem 275:8287–8289

Chang JY, Li L, Canals F, Aviles FX (2000b) The unfolding pathway and conformational stability of potato carboxypeptidase inhibitor. J Biol Chem 275:14205–14211

Charras GT, Horton MA (2002) Single cell mechanotransduction and its modulation analyzed by atomic force microscope indentation. Biophys J 82:2970–2981

Chen SJ, Dill KA (2000) RNA folding energy landscapes. Proc Natl Acad Sci USA 97:646–651

Chen L, Wildegger G, Kiefhaber T, Hodgson KO, Doniach S (1998) Kinetics of lysozyme refolding: structural characterization of a non-specifically collapsed state using time-resolved X-ray scattering. J Mol Biol 276:225–237

Chen J, Chen Y, Gong P, Jiang Y, Li YM, Zhao YF (2002a) Novel phosphoryl derivatization method for peptide sequencing by electrospray ionization mass spectrometry. Rapid Commun. Mass Spectrom 16:531–536

Chen L, Haushalter KA, Lieber CM, Verdine GL (2002b) Direct visualization of a DNA glycosylase searching for damage. Chem Biol 9:345–350

Cheng J, Wu L, Heller MJ, Sheldon E, Diver J, O'Connell JP, Smolko D, Jalali S, Willoughby D (2002) Integrated portable biological detection system. US Patent 6,403,367

Cherny DI, Jovin TM (2001) Electron and scanning force microscopy studies of alterations in supercoiled DNA tertiary structure. J Mol Biol 313:295–307

Cho SJ, Quinn AS, Stromer MH, Dash S, Cho J, Taatjes DJ, Jena BP (2002) Structure and dynamics of the fusion pore in live cells. Cell Biol Int 26:35–42

Choy WY, Mulder FA, Crowhurst KA, Muhandiram DR, Millett IS, Doniach S, Forman-Kay JD, Kay LE (2002) Distribution of molecular size within an unfolded state ensemble using small angle X-ray scattering and pulse field gradient NMR techniques. J Mol Biol 316:101–112

Christendat D, Yee A, Dharamsi A, Kluger Y, Gerstein M, Arrowsmith CH, Edwards AM (2000) Structural proteomics: prospects for high throughput sample preparation. Prog Biophys Mol Biol 73:3393–45

Clementi C, Jennings PA, Onuchic JN (2000a) How native-state topology affects the folding of dihydrofolate reductase and interleukin-1β. Proc Natl Acad Sci USA 97:5871–5876

Clementi C, Nymeyer H, Onuchic JN (2000b) Topological and energetic factors: what determines the structural details of the transition state ensemble and "en-route" intermediates for protein folding? An investigation for small globular proteins. J Mol Biol 298:937–953

Cobon GS, Verrills N, Papakostopoulos P, Eastwood H, Linnane AW (2002) The proteomics of aging. Biogerontology 3:133–136

Cohen MJ, Karasek FW (1970) Plasma chromatography – a new dimension for gas chromatography and mass spectrometry. J Chromatographic Sci 8:330–337

Cohen MJ, Crowe RW (1973) Apparatus and methods for detecting, separating, concentrating and measuring electronegative trace vapors. US Patent 3,742,213

Cohen MJ, Carroll DI, Wernlund RF, Kilpatrick WD (1972) Apparatus and methods for separating, concentrating, detecting, and measuring trace gases. US Patent 3,699,333

Coimbra MA, Goncalves F, Barros AS, Delgadillo I (2002) Fourier transform infrared spectroscopy and chemometric analysis of white wine polysaccharide extracts. J Agric Food Chem 50:3405–3411

Coligan JE, Dunn BM, Ploegh HL, Speicher DW, Wingfield PT (ed) (1996) Current protocols in protein science. Wiley & Sons, New York

Costantino HR, Griebenow K, Mishra P, Langer R, Klibanov AM (1995) Fourier transform infrared spectroscopic investigation of protein stability in the lyophilized form. Biochim Biophys Acta 1253:69–74

Crawford OH (1999) A fast, stochastic threading algorithm for proteins. Bioinformatics 15:66–71

Csermely P (1999) Chaperone-percolator model: a possible molecular mechanism of Anfinsen-cage-type chaperones. BioEssays 21:959–965

Cui XD, Primak A, Zarate X, Tomfohr J, Sankey OF, Moore AL, Moore TA, Gust D, Harris G, Lindsay SM (2001) Reproducible measurement of single-molecule conductivity. Science 294:571–574

Czaplewski C, Rodziewicz-Motowidlo S, Liwo A, Ripoll DR, Wawak RJ, Scheraga HA (2000) Molecular simulation study of cooperativity in hydrophobic association. Protein Sci 9:1235–1245

D'Alessio G (1999a) Evolution of oligomeric proteins – the unusual case of a dimeric ribonuclease. Eur J Biochem 266:699–708

D'Alesso G (1999b) The evolutionary transition from monomeric to oligomeric proteins: tools, the environment, hypotheses. Progress Biophys Mol Biol 72:271–298

Dammer U, Popescu O, Wagner P, Anselmetti D, Güntherodt HJ, Misevic GN (1995) Binding strength between cell adhesion proteoglycans measured by atomic force microscopy. Science 267:1173–1175

Dammer U, Hegner M, Anselmetti D, Wagner P, Dreier M, Huber W, Güntherodt HJ (1996) Specific antigen/antibody interactions measured by force microscopy. Biophys J 70:2437–2441

Davda J, Labhasetwar V (2002) Characterization of nanoparticle uptake by endothelial cells. Int J Pharm 233:51–59

Davies DK (1994) Pulsed ionization ion mobility sensor. US Patent 5,300,773

Dawson RMC, Elliott DC, Elliott WH, Jones KM (1969) Data for biochemical research. Oxford University Press, 2nd Ed,

de Cock H, Brandenburg K, Wiese A, Holst O, Seydel U (1999) Non-lamellar structure and negative charges of lipopolysaccharides required for efficient folding of outer membrane protein PhoE of *Escherichia coli*. J Biol Chem 274:5114–5119

Demers LM, Ginger DS, Park SJ, Li Z, Chung SW, Mirkin CA (2002) Direct patterning of modified oligonucleotides on metals and insulators by dip-pen nanolithography. Science 296:1836–1838

Dempsey BR, Economou A, Dunn SD, Shilton BH (2002) The ATPase domain of SecA can form a tetramer in solution. J Mol Biol 315:831–843

de Paris R, Strunz T, Oroszlan K, Güntherodt HJ, Hegner M (2000) Force spectroscopy and dynamics of the biotin-avidin bond studied by scanning force microscopy. Single Mol 1:285–290

Desmeules P, Grandbois M, Bondarenko VA, Yamazaki A, Salesse C (2002) Measurement of membrane binding between recoverin, a calcium-myristoyl switch protein, and lipid bilayers by AFM-based force spectroscopy. Biophys J 82:3343–3350

Dierksen K, Typke D, Hegerl R, Koster AJ, Baumeister W (1992) Towards automatic electron tomography. Ultramicroscopy 40:71–87

Dill KA, Fiebig KM, Chan HS (1993) Cooperativity in protein folding kinetics. Proc Natl Acad Sci USA 90:1942–1946

Ding FX, Schreiber D, VerBerkmoes NC, Becker JM, Naider F (2002) The chain length dependence of helix formation of the second transmembrane domain of a G protein-coupled receptor of *Saccharomyces cerevisiae*. J Biol Chem 277:14483–14492

Djuricic D, Hill HA, Lo KK, Wong LL (2002) A scanning tunneling microscopy (STM) investigation of complex formation between cytochrome $P450_{cam}$ and putidaredoxin. J Inorg Biochem 88:362–367

Dong A, Malecki JM, Lee L, Carpenter JF, Lee JC (2002) Ligand-induced conformational and structural dynamics changes in *Escherichia coli* cyclic AMP receptor protein. Biochemistry 41:6660–6667

Döring HR, Arnold G, Adler J, Röbel T, Riemenschneider J (1999) Photo-ionization ion mobility spectrometry. US Patent 5,968,837

Doyle R, Simons K, Qian H, Baker D (1997) Local interactions and the optimization of protein folding. Proteins 29:282–291

Drablos F (1999) Clustering of non-polar contacts in proteins. Bioinformatics 15:501–509

Duarte IF, Barros A, Delgadillo I, Almeida C, Gil AM (2002) Application of FTIR spectroscopy for the quantification of sugars in mango juice as a function of ripening. J Agric Food Chem 50:3104–3111

Duncan MD, Bashkansky M, Reintjes j (1998) Subsurface defect detection in materials using optical coherence tomography. Optics Express 2:540–545

Durbin SD, Carlson WE (1992) Lysozyme crystal growth studied by atomic force microscopy. J Cryst Growth 122:71–79

Durbin SD, Carson WE, Saros MT (1993) *In situ* studies of protein crystal growth by atomic force microscopy. J Phys D Appl Phys 26:B128–B132

Dwek MV, Rawlings SL (2002) Current perspectives in cancer proteomics. Mol Biotechnol 22:139–152

Dworzanski JP, McClennen WH, Cole PA, Thornton SN, Meuzelaar HLC, Arnold NS, Snyder AP (1997) Field-portable, automated pyrolysis/GC/IMS system for rapid biomarker detection in aerosols: a feasibility study. Field Anal Chem Technol 1:295–305

Dzwolak W, Kato M, Taniguchi Y (2002) Fourier transform infrared spectroscopy in high-pressure studies on proteins. Biochim Biophys Acta 1595:131–144

Edwards AM, Arrowsmith CH, des Pallieres B (2000) Proteomics: New tools for a new era. Modern Drug Discovery 5:35–44

Edwards AM, Kus B, Jansen R, Greenbaum D, Greenblatt J, Gerstein M (2002) Bridging structural biology and genomics: assessing protein interaction data with known complexes. Trends Genet 18:529–536

Efimov AV (1998) A structural tree for protein containing S-like β-sheets. FEBS Lett 437:246–250

Efimov AV (1999) Complementary packing of α-helices in proteins. FEBS Lett 463:3–6

Egawa A, Chiba N, Homma K, Chinone K, Muramatsu H (1999) High-speed scanning by dual feedback control in SNOM/AFM. J Microsc 194:325–328

Egea PF, Rochel N, Birck C, Vachette P, Timmins PA, Moras D (2001) Effects of ligand binding on the association properties and conformation in solution of retinoic acid receptors RXR and RAR. J Mol Biol 307:557–576

Eiceman GA, Karas Z (1994) Ion Mobility Spectrometry. CRC Press, Boca Raton

Eiceman GA, Ferris MJ, Anderson GK, Danen WC, Tiee JJ (1988) Laser desorption and ionization of solid polycyclic aromatic hydrocarbons in air with analysis by ion mobility spectrometry. Anal Lett 21:539–552

Eiceman GA, Tadjikov B, Krylov E, Nazarov EG, Miller RA, Westbrook J, Funk P (2001) Miniature radio-frequency mobility analyzer as a gas chromatographic detector for oxygen-containing volatile organic compounds, pheromones and other insect attractants. J Chromatogr A 917:205–217

El Khattabi M, Ockhuijsen C, Bitter W, Jaeger KE, Tommassen J (1999) Specifity of the lipase-specific foldases of gram-negative bacteria and the role of the membrane anchor. Mol Gen Genet 261:770–776

Ellis RJ, Hartl FU (1999) Principles of protein folding in the cellular environment. Curr Opin Struct Biol 9:102–110

Eve JK, Patel N, Luk SY, Ebbens SJ, Roberts CJ (2002) A study of single drug particle adhesion interactions using atomic force microscopy. Int J Pharm 238:17–27

Facci P, Alliata D, Cannistraro S (2001) Potential-induced resonant tunneling through a redox metalloprotein investigated by electrochemical scanning probe microscopy. Ultramicroscopy 89:291–298

Fagas G, Cuniberti G, Richter K (2002) Molecular wire-nanotube interfacial effects on electron transport. Ann N Y Acad Sci 960:216–224

Favier AL, Schoehn G, Jaquinod M, Harsi C, Chroboczek J (2002) Structural studies of human enteric adenovirus type 41. Virology 293:75–85

Fermi G, Perutz MF, Shaanan B, Fourme R (1984) The crystal structure of human deoxyheamoglobin at 1.74 Å resolution. J Mol Biol 175:159–174

Fernandez M, Keyrilainen J, Serimaa R, Torkkeli M, Karjalainen-Lindsberg ML, Tenhunen M, Thomlinson W, Urban V, Suortti P (2002) Small angle X-ray scattering studies of human breast tissue samples. Phys Med Biol 47:577–592

Fersht AR (1995a) Characterizing transition states in protein folding – an essential step in the puzzle. Curr Opin Struct Biol 5:79–84

Fersht AR (1995b) Optimization of rates of protein folding – the nucleation-condensation mechanism and its implications. Proc Natl Acad Sci USA 92:10869–10873

Fersht AR, Matouschek A, Serrano L (1992) The folding of an enzyme. 1. Theory of protein engineering analysis of stability and pathway of protein folding. J Mol Biol 224:771–782

Fetler L, Tauc P, Baker DP, Macol CP, Kantrowitz ER, Vachette P (2002) Replacement of Asp-162 by Ala prevents the cooperative transition by the substrates while enhancing the effect of the allosteric activator ATP on E. coli aspartate transcarbamoylase. Protein Sci 11:1074–1081

Fetterolf DD, Clark TD (1993) Detection of trace explosive evidence by ion mobility spectrometry. J Forensic Sci 38:28–39

Feughelman M, Lyman DJ, Willis BK (2002) The parallel helices of the intermediate filaments of α-keratin. Int J Biol Macromol 30:95–96

Figeys D (2002a) Proteomics approaches in drug discovery. Anal Chem 74:412A–419A

Figeys D (2002b) Adapting arrays and lab-on-a-chip technology for proteomics. Proteomics 2:373–382

Figeys D (2002c) Functional proteomics: mapping protein-protein interactions and pathways. Curr Opin Mol Ther 4:210–215

Florin EL, Moy VT, Gaub HE (1994) Adhesion forces between individual ligand-receptor pairs. Science 264:415–417

Florin EL, Pralle A, Horber JK, Stelzer EH (1997) Photonic force microscope based on optical tweezers and two-photon excitation for biological applications. J Struct Biol 119:202–211

Forge V, Hoshino M, Kuwata K, Arai M, Kuwajima K, Batt CA, Goto Y (2000) Is folding of β-lactoglobulin non-hierarchic? Intermediate with native-like β-sheet and non-native α-helix. J Mol Biol 296:1039–1051

Freeman R, Goodfellow M, Gould FK, Hudson SJ, Lightfoot NF (1990) Pyrolysis mass spectrometry (Py-MS) for the rapid epidemiological typing of clinically significant bacterial pathogens. J Med Microbiol 32:283–286

Freeman R, Sisson PR, Heatherington CS (1995) Pyrolysis mass spectrometry. Methods Mol Biol 46:97–105

Freeman R, Sisson PR, Barer MR, Ward AC, Lightfoot NF (1997) A highly discriminatory method for the direct comparison of two closely related bacterial populations by pyrolysis mass spectrometry. Zentralbl Bakteriol 285:285–290

Frey AH (1961) Auditory system response to modulated electromagnetic energy. Aerospace Med 32:1140–1142

Frey AH (1962) Human auditory system response to modulated electromagnetic energy. J Appl Physiol 17:689–692

Frey AH (1993) Electromagnetic field interactions with biological systems. FASEB Journal 7:272–281

Frey AH, Messener R (1973) Human perception of illumination with pulsed UHF electromagnetic energy. Science 181:356–358

Frey AH, Corin E (1979) Holographic assessment of a hypothesized microwave hearing mechanism. Science 206:232–234

Frey TG, Mannella CA (2000) The internal structure of mitrochondria. Trends Biochem Sci 25:319–324

Fringeli UP, Goette J, Reiter G, Siam M, Baurecht D (1998) Structural investigation of oriented membrane assemblies by FTIR-ATR spectroscopy. AIP Conf Proc 430:729–747

Fritz J, Baller MK, Lang HP, Rothuizen H, Vettiger P, Meyer E, Güntherodt HJ, Gerber C, Gimzewski JK (2000) Translating biomolecular recognition into nanomechanics. Science 288:316–318

Frolow F, Kalb AJ, Yariv J (1994) Structure of a unique, twofold symmetrical haem-binding site. Nature Struct Biol 1:453–460

Furuike S, Ito T, Yamazaki M (2001) Mechanical unfolding of single filamin A (ABP-280) molecules detected by atomic force microscopy. FEBS Lett 498:72–75

Gaietta G, Deerinck TJ, Adams SR, Bouwer J, Tour O, Laird DW, Sosinsky GE, Tsien RY, Ellisman MH (2002) Multicolor and electron microscopic imaging of connexin trafficking. Science 296:503–507

Gaigneaux A, Ruysschaert JM, Goormaghtigh E (2002) Infrared spectroscopy as a tool for discrimination between sensitive and multiresistant K562 cells. Eur J Biochem 269:1968–1973

Gall J (1994) Method and system for altering consciousness. US Patent 5,289,438; and references therein

Gallagher T, Alexander P, Bryan P, Gilliland GL (1994) Two crystal structures of the B1 immunoglobin-binding domain of streptococcal protein G and comparison with NMR. Biochemistry 33:4721–4729

Gallardo K, Job C, Groot SP, Puype M, Demol H, Vandekerckhove J, Job D (2002) Proteomics of *Arabidopsis* seed germination. A comparative study of wild-type and gibberellin-deficient seeds. Plant Physiol 129:823–837

Galzitskaya OV, Surin AK, Nakamura H (2000) Optimal region of average sidechain entropy for fast protein folding. Protein Sci 9:580–586

Gao H, Oberringer M, Englisch A, Hanselmann RG, Hartmann U (2001) The scanning near-field optical microscope as a tool for proteomics. Ultramicroscopy 86:145–150

Garcia P, Serrano L, Durand D, Rico M, Bruix M (2001) NMR and SAXS characterization of the denatured state of the chemotactic protein CheY: implications for protein folding initiation. Protein Sci 10:1100–1112

Garcia-Hernandez E, Hernandez-Arana A (1999) Structural basis of lectin-carbohydrate affinities: comparison with protein-folding energetics. Protein Sci 8:1075–1086

Gaub HE, Fernandez JM (1998) The molecular elasticity of individual proteins studied by AFM-related techniques. AvH Magazin 71:11–18

Genick UK, Borgstahl GEO, Ng K, Ren Z, Pradervand C, Burke PM, Srajer V, Teng TY, Schildkamp W, McRee DE, Moffat K, Getzoff ED (1997) Structure of a protein photocycle intermediate by millisecond time-resolved crystallography. Science 275:1471–1475

Gera JF, Hazbun TR, Fields S (2002) Array-based methods for identifying protein-protein and protein-nucleic acid interactions. Methods Enzymol 350:499–512

Ghirlanda G, Lear JD, Ogihara NL, Eisenberg D, DeGrado WF (2002) A hierarchic approach to the design of hexameric helical barrels. J Mol Biol 319:243–253

Giessibl FJ, Hembacher S, Bielefeldt H, Mannhart J (2000) Subatomic features on the Silicon(111)-(7x7) surface observed by atomic force microscopy. Science 289:422–426

Gilardi G, Fantuzzi A, Sadeghi SJ (2001) Engineering and design in the bioelectrochemistry of metalloproteins. Curr Opin Struct Biol 11:491–499

Goebel J, Breit U (2000) Ion mobility spectrometer. US Patent 6,049,076

Goldbeck RA, Thomas YG, Chen E, Esquerra RM, Kliger DS (1999) Multiple pathways on a protein-folding energy landscape: kinetic evidence. Proc Natl Acad Sci USA 96:2782–2787

Goldsbury C, Aebi U, Frey P (2001) Visualizing the growth of Alzheimer's A β-amyloid-like fibrils. Trends Mol Med 7:582

Goodacre R (1994) Characterisation and quantification of microbial systems using pyrolysis mass spectrometry: introducing neural networks to analytical pyrolysis. Microbiology Europe 2:16–22

Goodacre R, Kell DB (1996) Pyrolysis mass spectrometry and its applications in biotechnology. Curr Opin Biotechnol 7:20–28

Goodacre R, Howell SA, Noble WC, Neal MJ (1996) Sub-species discrimination, using pyrolysis mass spectrometry and self-organising neural networks, of *Propionibacterium acnes* isolated from normal human skin. Zentralbl Bakteriol 284:501–515

Goodacre R, Rooney PJ, Kell DB (1998a) Discrimination between methicillin-resistant and methicillin-susceptible *Staphylococcus aureus* using pyrolysis mass spectrometry and artificial neural networks. J Antimicrob Chemother 41:27–34

Goodacre R, Timmins EM, Burton R, Kaderbhai N, Woodward AM, Kell DB, Rooney PJ (1998b) Rapid identification of urinary tract infection bacteria using hyperspectral whole-organism fingerprinting and artificial neural networks. Microbiology 144:1157–1170

Goodacre R, Shann B, Gilbert RJ, Timmins EM, McGovern AC, Alsberg BK, Kell DB, Logan NA (2000) Detection of the dipicolinic acid biomarker in *Bacillus* spores using Curie point pyrolysis mass spectrometry and Fourier transform infrared spectroscopy. Anal Chem 72:119–127

Goodfellow M, Freeman R, Sisson PR (1997) Curie point pyrolysis mass spectrometry as a tool in clinical microbiology. Zentralbl Bakteriol 285:133–156

Gordon L, Mobley PW, Pilpa R, Sherman MA, Waring AJ (2002) Conformational mapping of the N-terminal peptide of HIV-1 gp41 in membrane environments using ^{13}C-enhanced Fourier transform infrared spectroscopy. Biochim Biophys Acta 1559:96–120

Goto Y, Aimoto S (1991) Anion and pH-dependent conformational transition of an amphiphilic polypeptide. J Mol Biol 218:387–396

Goto Y, Hoshino M, Kuwata K, Batt CA (1999) Folding of β-lactoglobulin, a case of the inconsistency of local and non-local interactions. In: Kuwajima K and Arai M (ed) Old and New Views of Protein Folding. Elsevier, Amsterdam, 3–11

Govindarajan S, Goldstein RA (1995) Optimal local propensities for model proteins. Proteins 22:413–418

Grabner B, Landis WJ, Roschger P, Rinnerthaler S, Peterlik H, Klaushofer K, Fratzl P (2001) Age- and genotype-dependence of bone material properties in the osteogenesis imperfecta murine model (oim). Bone 29:453–457

Grandori R, Matecko I, Müller N (2002) Uncoupled analysis of secondary and tertiary protein structure by circular dichroism and electrospray ionization mass spectrometry. J Mass Spectrom 37:191–196

Grantcharova VP, Riddle DS, Baker D (2000) Long-range order in the src SH3 folding transition state. Proc Natl Acad Sci USA 97:7084–7089

Grayhack EJ, Phizicky EM (2001) Genomic analysis of biochemical function. Curr Opin Chem Biol 5:34–39

Griffin TJ, Goodlett DR, Aebersold R (2001) Advances in proteome analysis by mass spectrometry. Curr Opin Biotechnol 12:607–612

Grigoriev IV, Rakhmaninova AB, Mironov AA (1998) Simulated annealing for α-helical protein folding: searches in vicinity of the "molten globule" state. J Biomol Struct Dyn 16:115–121

Grigoriev IV, Mironov AA, Rakhmaninova AB (1999) Refinement of helix boundaries in α-helical globular proteins (in Russian). Mol Biol (Moscow) 33:206–214

Griko YV (2000) Energetic basis of structural stability in the molten globule state of β-lactalbumin. J Mol Biol 297:1259–1268

Gromiha MM, Selvaraj S (1997) Influence of medium and long range interactions in different structural classes of globular proteins. J Biol Phys 23:151–162

Gromiha MM, Selvaraj S (1999) Importance of long-range interactions in protein folding. Biophys Chem 77:49–68

Gross M (1996) Linguistic analysis of protein folding. FEBS Lett 390:249–252

Gruebele M (1999) The fast protein-folding problem. Annu Rev Phys Chem 50:485–516

Gunning AP, Wilde PJ, Clark DC, Morris VJ, Parker ML, Gunning PA (1996) Atomic force microscopy of interfacial protein films. J Colloid Interface Sci 183:600–602

Gursky O (1999) Probing the conformation of a human apolipoprotein C-1 by amino acid substitutions and trimethylamine-N-oxide. Prot Sci 8:2055–2064

Gursky O, Alehkov S (2000) Temperature-dependent β-sheet formation in β-amyloid Aβ(1–40) peptide in water: uncoupling β-structure folding from aggregation. Biochim Biophys Acta 1476:93–102

Gutsche I, Holzinger J, Rößle M, Heumann H, Baumeister W, May RP (2000a) Conformational rearrangements of an archaeal chaperonin upon ATPase cycling. Curr Biol 10:405–408

Gutsche I, Mihalache O, Hegerl R, Typke D, Baumeister W (2000b) ATPase cycle controls the conformation of an archaeal chaperonin as visulaized by cryo-electron microscopy. FEBS Lett 477:278–282

Haas J, Lehr CM (2002) Developments in the area of bioadhesive drug delivery systems. Expert. Opin Biol Ther 2:287–298

Haider M, Uhlemann S, Schwan E, Rose H, Kabius B, Urban K (1998) Electron microscopy image enhanced. Nature 392:768–769

Hamada D, Kuroda Y, Tanaka T, Goto Y (1995) High helical propensity of the peptide fragments derived from β-lactoglobulin, a predominantly β-sheet protein. J Mol Biol 254:737–746

Hansma PK, Drake B, Marti O, Gould SAC, Prater CB (1989) The scanning ion-conductance microscope. Science Reports 243:641–643

Hardesty B, Tsalkova T, Kramer G (1999) Co-translational folding. Curr Opin Struct Biol 9:111–114

Häusler G, Lindner MW (1998) "Coherence radar" and "Spectral radar"–New tools for dermatological diagnosis. J Biomed Opt 3:21–31

Heimel J, Fischer UC, Fuchs H (2001) SNOM/STM using a tetrahedral tip and a sensitive current-to-voltage converter. J Microsc 202:53–59

Heinemann U, Frevert J, Hofmann K, Illing G, Maurer C, Oschkinat H, Saenger W (2000) An integrated approach to structural genomics. Prog Biophys Mol Biol 73:347–362

Heinemann U, Illing G, Oschkinat H (2001) High-throughput three-dimensional protein structure determination. Curr Opin Biotechnol 12:348–354

Helyer RJ, Kelley T, Berkeley RC (1997) Pyrolysis mass spectrometry studies on *Bacillus anthracis*, *Bacillus cereus* and their close relatives. Zentralbl Bakteriol 285:319–328

Henderson R (1996) Lecture "Resolution limits of microscopes", Laboratory of Molecular Biology, Cambridge

Henderson E, Haydon PG, Sakaguchi DS (1992) Actin filament dynamics in living glial cells imaged by atomic force microscopy. Science 257:1944–1946

Hertadi R, Ikai A (2002) Unfolding mechanics of *holo-* and *apo*calmodulin studied by the atomic force microscope. Protein Sci 11:1532–1538

Heyes CD, Wang J, Sanii LS, El-Sayed MA (2002) Fourier transform infrared study of the effect of different cations on bacteriorhodopsin protein thermal stability. Biophys J 82:1598–1606

Hilario J, Kubelka J, Syud FA, Gellman SH, Keiderling TA (2002) Spectroscopic characterization of selected β-sheet hairpin models. Biopolymers 67:233–236

Hildebrand G, Kunze S, Driver M (2001) Blood cell adhesion on sensor materials studied by light, scanning electron, and atomic-force microscopy. Ann Biomed Eng 29:1100–1105

Hill HH, Siems WF, St Louis RW, McMinn DG (1990) Ion mobility spectrometry. Anal Chem 62:1201–1209

Hodneland CD, Lee YS, Min DH, Mrksich M (2002) Selective immobilization of proteins to self-assembled monolayers presenting active site-directed capture ligands. Proc Natl Acad Sci USA 99:5048–5052

Honda S, Kobayashi N, Munekata E, Uedaira H (1999) Fragment reconstitution of a small protein: folding energetics of the reconstituted immunoglobulin binding domain B1 of stretococcal protein G. Biochemistry 38:1203–1213

Honda S, Kobayashi N, Munekata E (2000) Thermodynamics of a β-hairpin structure: evidence for cooperative formation of folding nucleus. J Mol Biol 295:269–278

Hoover DM, Ludwig ML (1997) A flavodoxin that is required for enzyme activation: the structure of oxidized flavodoxin from *Escherichia coli* at 1.8 Å resolution. Protein Sci 6:2525–2537

Horiuchi Y, Yagi K, Hosokawa T, Yamamoto N, Muramatsu H, Fujihira M (1999) Imaging of various surface properties of fluorescently labelled phospholipid Langmuir-Blodgett films with a combined scanning probe microscope. J Microsc 194:467–471

Hornemann S, Glockshuber R (1998) A scrapie-like unfolding intermediate of the prion protein domain PrP(121–231) induced by acidic pH. Proc Natl Acad Sci USA 95:6010–6014

Howald L, Lüthl R, Meyer E, Güntherodt HJ (1995) Atomic-force microscopy on the Si(111)-(7x7) surface. Phys Rev B Condens Matter 51:5484–5487

Huang CY, Getahun Z, Zhu Y, Klemke JW, DeGrado WF, Gai F (2002) Helix formation via conformation diffusion search. Proc Natl Acad Sci USA 99:2788–2793

Hubbard MJ (2002) Functional proteomics: The goalposts are moving. Proteomics 2:1069–1078

Hung K, Sun X, Ding H, Kalafatis M, Simioni P, Guo B (2002) A matrix-assisted laser desorption/ionization time-of-flight based method for screening the 1691 G ← A mutation in the factor V gene. Blood Coagul Fibrinolysis 13:117–122

Igartua M, Saulnier P, Heurtault B, Pech B, Proust JE, Pedraz JL, Benoit JP (2002) Development and characterization of solid lipid nanoparticles loaded with magnetite. Int J Pharm 233:149–157

Ikarashi Y, Itoh K, Maruyama Y (1991) Application of FRIT fast atom bombardment liquid chromatography / mass spectrometry for the determination of acetylcholine levels in rat brain regions. Biol Mass Spectrom. 20:21–25

Ikeda S, Morris VJ (2002) Fine-stranded and particulate aggregates of heat-denatured whey proteins visualized by atomic force microscopy. Biomacromolecules 3:382–389

Irbäck A, Peterson C, Potthast F, Sandelin E (1999) Design of sequences with good folding properties in coarse-grained protein models. Structure Fold Des 7:347–360

Ironside JW (1998) Prion diseases in man. J Pathol 186:227–234

Ishizawa F, Misawa S (1990) Capillary column pyrolysis - gas chromatography of hair: a short study in personal identification. J Forensic Sci Soc 30:201–209

Ito Y, Bleloch AL, Brown LM (1998) Nanofabrication of solid-state Fresnel lenses for electron optics. Nature 394:49–52

Ito T, Ota K, Kubota H, Yamaguchi Y, Chiba T, Sakuraba K, Yoshida M (2002) Roles for the two-hybrid system in exploration of the yeast protein interactome. Mol Cell Proteomics 1:561–566

Itoh H, Ogura M, Komatsuda A, Wakui H, Miura AB, Tashima Y (1999) A novel chaperone-activity-reducing mechanism of the 90-kDa molecular chaperone HSP90. Biochem J 343:697–703

Itzhaki LS, Otzen DE, Fersht AR (1995) The structure of the transition state for folding of chymotrypsin inhibitor 2 analysed by protein engineering methods: evidence for a nucleation-condensation mechanism for protein folding. J Mol Biol 254:260–288

Iverson TM, Luna-Chavez C, Cecchini G, Rees DC (1999) Structure of the *E. coli* fumarate reductase respiratory complex. Science 284:1961–1966

Jackson SE (1998) How do small single-domain proteins fold? Fold Des 3:R81–R91

Jäger D, Jungblut PR, Müller-Werdan U (2002) Separation and identification of human heart proteins. J Chromatogr B Analyt Technol Biomed Life Sci 771:131–153

Jain KK (2002) Recent advances in oncoproteomics. Curr Opin Mol Ther 4:203–209

Jeney S, Florin EL, Horber JK (2001) Use of photonic force microscopy to study single-motor-molecule mechanics. Methods Mol Biol 164:91–108

Jésior JC, Filhol A, Tranqui D (1994) FoldIt (light) – an interactive program for Macintosh computers to analyze and display Protein Data Bank coordinate files. J Appl Cryst 27:1075

Jésior JC (2000) Hydrophilic frameworks in proteins? J Protein Chem 19:93–103

Jiang M, Nölting B, Stayton PS, Sligar SG (1996) Surface-linked molecular monolayers of an engineered myoglobin: structure, stability, and function. Langmuier 12:1278–1283

Jiao Y, Cherny DI, Heim G, Jovin TM, Schaffer TE (2001) Dynamic interactions of p53 with DNA in solution by time-lapse atomic force microscopy. J Mol Biol 314:233–243

Jimenez CR, Eyman M, Lavina ZS, Gioio A, Li KW, van der Schors RC, Geraerts WP, Giuditta A, Kaplan BB, van Minnen J (2002) Protein synthesis in synaptosomes: a proteomics analysis. J Neurochem 81:735–744

Kaji N, Ueda M, Baba Y (2001) Direct measurement of conformational changes on DNA molecule intercalating with a fluorescence dye in an electrophoretic buffer solution by means of atomic force microscopy. Electrophoresis 22:3357–3364

Kamatari YO, Ohji S, Konno T, Seki Y, Soda K, Kataoka M, Akasaka K (1999) The compact and expanded denatured conformations of apomyoglobin in the methanol-water solvent. Protein Sci 8:873–882

Kandori H, Shimono K, Shichida Y, Kamo N (2002) Interaction of Asn105 with the retinal chromophore during photoisomerization of pharaonis phoborhodopsin. Biochemistry 41:4554–4559

Karasek FW (1970) Plasma chromatograph. Research/Development 21:34–37

Karl M (1994) Ion mobility spectrometer drift chamber. US Patent 5,280,175

Karplus M, Weaver DL (1994) Protein folding dynamics: the diffusion-collision model and experimental data. Protein Sci 3:650–668

Katakuse I, Matsuo T, Matsuda H, Shimonishi Y, Hong YM, Izumi Y (1982) Sequence determination of a peptide with 55 amino acid residues by Edman degradation and field desorption mass spectrometry. Biomed Mass Spectrom 9:64–68

Katou H, Hoshino M, Kamikubo H, Batt CA, Goto Y (2001) Native-like β-hairpin retained in the cold-denatured state of bovine β-lactoglobulin. J Mol Biol 310:471–484

Kawata Y, Kawagoe M, Hongo K, Mikuya T, Higurashi T, Mizobata T, Nagai J (1999) Functional communications between the apical and equatorial domains of GroEL through the intermediate domain. Biochemistry 38:15731–15740

Keller RA (1975) Plasma chromatograph, an atmospheric pressure chemical ionization drift-time spectrometer. Am Lab 7:35–44

Keller T, Miki P, Regenscheit P, Dirnhofer R, Schneider A, Tsuchihashi H (1998) Detection of designer drugs in human hair by ion mobility spectrometry. Forensic Sci Int 94:55–63

Kellermayer MS, Smith SB, Bustamante C, Granzier HL (2001) Mechanical fatigue in repetitively stretched single molecules of titin. Biophys J 80:852–863

Kendrew JC, Dickerson RE, Strandberg BE, Hart RJ, Davies DR, Phillips DC, Shore VC (1960) Structure of myoglobin: a three-dimensional Fourier synthesis at 2 Å resolution. Nature 185:422–427

Kenyon RG, Ferguson EV, Ward AC (1997) Application of neural networks to the analysis of pyrolysis mass spectra. Zentralbl Bakteriol 285:267–277

Kersten B, Burkle L, Kuhn EJ, Giavalisco P, Konthur Z, Lueking A, Walter G, Eickhoff H, Schneider U (2002) Large-scale plant proteomics. Plant Mol Biol 48:133–141

Kharakoz DP (1989) Volumetric properties of proteins and their analogs in diluted water solutions. 1. Partial volumes of amino acids at 15–55 °C. Biophys Chem 34:115–125

Kharakoz DP (1991) Volumetric properties of proteins and their analogs in diluted water solutions. 2. Partial adiabatic compressibilities of amino acids at 15–70 °C. J Phys Chem 95:5634–5642

Kharakoz DP (1997) Partial volumes and compressibilities of extended polypeptide chains in aqueous solution: additivity scheme and implication of protein unfolding at normal and high pressure. Biochemistry 36:10276–10285

Khomutov GB, Belovolova LV, Gubin SP, Khanin VV, Obydenov AY, Sergeev-Cherenkov AN, Soldatov ES, Trifonov AS (2002) STM study of morphology and electron transport features in cytochrome c and nanocluster molecule monolayers. Bioelectrochemistry 55:177–181

Kienzl E, Jellinger K, Stachelberger H, Linert W (1999) Iron as catalyst for oxidative stress in the pathogenesis of Parkinson's disease? Life Sci 65:1973–1976

Kim Y, Prestegard JH (1990) Refinement of the NMR structures for acyl carrier protein with scalar coupling data. Proteins 8:377–385

Kim JM, Ohtani T, Sugiyama S, Hirose T, Muramatsu H (2001) Simultaneous topographic and fluorescence imaging of single DNA molecules for DNA analysis with a scanning near-field optical/atomic force microscope. Anal Chem 73:5984–5991

Kinney JH, Pople JA, Marshall GW, Marshall SJ (2001) Collagen orientation and crystallite size in human dentin: a small angle X-ray scattering study. Calcif Tissue Int 69:31–37

Kintz P, Cirimele V, Sengler C, Mangin P (1995) Testing human hair and urine for anhydroecgonine methyl ester, a pyrolysis product of cocaine. J Anal Toxicol 19:479–482

Kizil R, Irudayaraj J, Seetharaman K (2002) Characterization of irradiated starches by using FT-Raman and FTIR spectroscopy. J Agric Food Chem 50:3912–3918

Klade CS (2002) Proteomics approaches towards antigen discovery and vaccine development. Curr Opin Mol Ther 4:216–223

Kline AD, Braun W, Wüthrich K (1988) Determination of the complete three-dimensional structure of the α-amylase inhibitor tendamistat in aqueous solution by nuclear magnetic resonance and distance geometry. J Mol Biol 204:675–724

Kneipp J, Beekes M, Lasch P, Naumann D (2002) Molecular changes of preclinical scrapie can be detected by infrared spectroscopy. J Neurosci 22:2989–2997

Koga N, Takada S (2001) Roles of native topology and chain-length scaling in protein folding: a simulation study with a Go-like model. J Mol Biol 313:171–180

Kohno M, Enatsu M, Yoshiizumi M, Kugimiya W (1999) High-level expression of *Rhizopus niveus* lipase in the yeast *Saccharomyces cerevisiae* and structural properties of the expressed enzyme. Protein Expr Purif 15:327–335

Kojima M, Tanokura M, Maeda M, Kimura K, Amemiya Y, Kihara H, Takahashi K (2000) pH-dependent unfolding of aspergillopepsin II studied by small angle X-ray scattering. Biochemistry 39:1364–1372

Konan YN, Gurny R, Allemann E (2002) Preparation and characterization of sterile and freeze-dried sub-200 nm nanoparticles. Int J Pharm 233:239–252

Koradi R, Billeter M, Wüthrich K (1996) MOLMOL: a program for display and analysis of macromolecular structures. J Mol Graphics 14:51–55

Korchev YE, Bashford CL, Milovanovic M, Vodyanoy I, Lab MJ (1997) Scanning ion conductance microscopy of living cells. Biophys J 73:653–658

Korchev YE, Gorelik J, Lab MJ, Sviderskaya EV, Johnston CL, Coombes CR, Vodyanoy I, Edwards CR (2000a) Cell volume measurement using scanning ion conductance microscopy. Biophys J 78:451–457

Korchev YE, Raval M, Lab MJ, Gorelik J, Edwards CR, Rayment T, Klenerman D (2000b) Hybrid scanning ion conductance and scanning near-field optical microscopy for the study of living cells. Biophys J 78:2675–2679

Koshland DE Jr, Hamadani K (2002) Proteomics and models for enzyme cooperativity. J Biol Chem 277:46841–46844

Köster H (2001a) DNA sequencing by mass spectrometry. US Patent 6,194,144

Köster H (2001b) DNA sequencing by mass spectrometry. US Patent 6,225,450

Kotiaho T, Lauritsen FR, Degn H, Paakkanen H (1995) Membrane inlet ion mobility spectrometer for on-line measurement of ethanol in beer and in yeast fermentation. Anal Chim Acta 309:317–325

Kramer G, Kudlicki W, McCarthy D, Tsalkova T, Simmons D, Hardesty B (1999) N-terminal and C-terminal modifications affect folding, release from the ribosomes and stability of *in vitro* synthesed proteins. Int J Biochem Cell Biol 31:231–241

Kraulis PJ (1991) MOLSRCIPT – a program to produce both detailed and schematic plots of protein structures. J Appl Crystallogr 24:946–950

Krautbauer R, Pope LH, Schrader TE, Allen S, Gaub H (2002) Discriminating small molecule DNA binding modes by single molecule force spectroscopy. FEBS Lett 510:154–158

Kukar T, Eckenrode S, Gu Y, Lian W, Megginson M, She JX, Wu D (2002) Protein microarrays to detect protein-protein interactions using red and green fluorescent proteins. Anal Biochem 306:50–54

Kuwajima K, Yamaya H, Sugai S (1996) The burst-phase intermediate in the refolding of β-lactoglobulin studied by stopped-flow circular dichroism and absorption spectroscopy. J Mol Biol 264:806–822

Kuznetsov YG, Malkin AJ, Lucas RW, McPherson A (2000) Atomic force microscopy studies of icosahedral virus crystal growth. Colloids Surf B Biointerfaces 19:333–346

Lasch P, Pacifico A, Diem M (2002) Spatially resolved IR microspectroscopy of single cells. Biopolymers 67:335–338

Laurell T, Marko-Varga G (2002) Miniaturisation is mandatory unravelling the human proteome. Proteomics 2:345–351

Lawrence AH, Barbour RJ, Sutcliffe R (1991) Identification of wood species by ion mobility spectrometry. Anal Chem 63:1217–1221

Leahy DJ, Hendrickson WA, Aukhil I, Erickson HP (1992) Structure of a fibronectin type III domain from tenascin phased by MAD analysis of the selenomethionyl protein. Science 258:987–991

Leasure CS, Fleischer ME, Anderson GK, Eiceman GA (1986) Photoionization in air with ion mobility spectrometry using a hydrogen discharge lamp. Anal Chem 58:2142–2147

Leaves NI, Sisson PR, Freeman R, Jordens JZ (1997) Pyrolysis mass spectrometry in epidemiological and population genetic studies of *Haemophilus influenzae*. J Med Microbiol 46:204–207

Lee DS, Wu C, Hill HH (1998) Detection of carbohydrates by electrospray ionization / ion mobility spectrometry following microbore high-performance liquid chromatography. J Chromatogr 822:1–9

Lee KA, Craven KB, Niemi GA, Hurley JB (2002a) Mass spectrometric analysis of the kinetics of in vivo rhodopsin phosphorylation. Protein Sci 11:862–874

Lee KB, Park SJ, Mirkin CA, Smith JC, Mrksich M (2002b) Protein nanoarrays generated by dip-pen nanolithography. Science 295:1702–1705

Lee SW, Mao CB, Flynn CE, Belcher AM (2002c) Ordering of quantum dots using genetically engineered viruses. Science 296:892–895

Leonhardt JW (1996) New detectors in environmental monitoring using tritium sources. J Radioanlyt Nucl Chem 206:333–339

Leonhardt JW, Rohrbeck W, Bensch H (2001) A high resolution IMS for environmental studies. Supplement to the catalogue for the IMS supplied by the IUT Institute for Environmental Technologies Ltd, Berlin

Li T, Talvenheimo J, Zeni L, Rosenfeld R, Stearns G, Arakawa T (2002) Changes in protein conformation and dynamics upon complex formation of brain-derived neurotrophic factor and its receptor: investigation by isotope-edited Fourier transform IR spectroscopy. Biopolymers 67:10–19

Lim SO, Park SJ, Kim W, Park SG, Kim HJ, Kim YI, Sohn TS, Noh JH, Jung G (2002) Proteome analysis of hepatocellular carcinoma. Biochem Biophys Res Commun 291:1031–1037

Lin JC (1978) Microwave Auditory Effects and Applications. Charles C Thomas, Publisher, Springfield, IL, USA

Lin JC (1989) Electromagnetic Interaction with Biological Systems. Plenum Press, New York

Lin JC (ed) (2000) Advances in Electromagnetic Fields in Living Systems. Vol 3. Kluwer/Plenum, New York

Lin H, Cornish VW (2002) Screening and selection methods for large-scale analysis of protein function. Angew Chem Int Ed Engl 2002 41:4402–4425

Linderoth NA, Simon MN, Russel M (1997) The filamentous phage pIV multimer visualized by scanning transmission electron microscopy. Science 278:1635–1638

Lindqvist M, Graslund A (2001) An FTIR and CD study of the structural effects of G-tract length and sequence context on DNA conformation in solution. J Mol Biol 314:423–432

Lindsay SM, Thundat T, Nagahara L, Knipping U, Rill RL (1989) Images of DNA double helix in water. Science 244:1063–1064

Liphardt J, Onoa B, Smith SB, Tinoco I Jr, Bustamante C (2001) Reversible unfolding of single RNA molecules by mechanical force. Science 292:733–737

Liu M, Barth A (2002) Mapping nucleotide binding site of calcium ATPase with IR spectroscopy: effects of ATP γ-phosphate binding. Biopolymers 67:267–270

Liu H, Berger SJ, Chakraborty AB, Plumb RS, Cohen SA (2002) Multidimensional chromatography coupled to electrospray ionization time-of-flight mass spectrometry as an alternative to two-dimensional gels for the identification and analysis of complex mixtures of intact proteins. J Chromatogr B Analyt Technol Biomed Life Sci 782:267–289

Lowery OM (1992) Silent subliminal presentation system. US Patent 5,159,703

Lubman DM, Kronick MN (1982) Plasma chromatography with laser-produced ions. Anal Chem 54:1546–1551

Lubman DM, Kronick MN (1983) Multiwavelength-selective ionization of organic compounds in an ion mobility spectrometer. Anal Chem 55:867–873

MacBeath G (2002) Protein microarrays and proteomics. Nat Genet 32 Suppl 2:526–532

Magee JG, Goodfellow M, Sisson PR, Freeman R, Lightfoot NF (1997) Differentiation of *Mycobacterium senegalense* from related non-chromogenic mycobacteria using pyrolysis mass spectrometry. Zentralbl Bakteriol 285:278–284

Malins DC, Hellstrom KE, Anderson KM, Johnson PM, Vinson MA (2002) Antioxidant-induced changes in oxidized DNA. Proc Natl Acad Sci USA 99:5937–5941

Malkin AJ, Land TA, Kuznetsov YG, McPherson A, DeYoreo JJ (1995) Investigation of virus crystal growth mechanisms by *in situ* atomic force microscopy. Phys Rev Lett 75:2778–2781

Malkin AJ, Plomp M, McPherson A (2002) Application of atomic force microscopy to studies of surface processes in virus crystallization and structural biology. Acta Crystallogr D Biol Crystallogr 58:1617–1621

Man WJ, White IR, Bryant D, Bugelski P, Camilleri P, Cutler P, Heald G, Lord PG, Wood J, Kramer K (2002) Protein expression analysis of drug-mediated hepatotoxicity in the *Sprague-Dawley* rat. Proteomics 2:1577–1585

Maret W, Heffron G, Hill HA, Djuricic D, Jiang LJ, Vallee BL (2002) The ATP/metallothionein interaction: NMR and STM. Biochemistry 41:1689–1694

Marple VA, Chein CM (1980) Virtual impactors: a theroretical study. Environmental Sci Technol 14:976–985

Marple VA, Olson BA, Miller NC (1998) The role of inertial particle collectors in evaluating pharmaceutical aerosol delivery systems. J Aerosol Med 11 Suppl 1:S139–S153

Martin SJ, Butler MA, Frye GC, Schubert WK (1998) Ion mobility spectrometer using frequency-domain separation. US Patent 5,789,745

Martzen MR, McCraith SM, Spinelli SL, Torres FM, Fields S, Grayhack EJ, Phizicky EM (1999) A biochemical genomics approach for identifying genes by the activity of their products. Science 286:1153–1155

Maruyama T, Nakajima M, Ichikawa S, Sano Y, Nabetani H, Furusaki S, Seki M (2001) Small angle X-ray scattering analysis of stearic acid modified lipase. Biosci Biotechnol Biochem 65:1003–1006

Mathur AB, Collinsworth AM, Reichert WM, Kraus WE, Truskey GA (2001) Endothelial, cardiac muscle and skeletal muscle exhibit different viscous and elastic properties as determined by atomic force microscopy. J Biomech 34:1545–1553

Matsko N, Klinov D, Manykin A, Demin V, Klimenko S (2001) Atomic force microscopy analysis of bacteriophages ΦKZ and T4. J Electron Microsc (Tokyo) 50:417–422

Matsuda H (1976) Double focusing mass spectrometers of second order. Atomic Masses and Fundamental Constants 5:185–191

Matsuda H (1981) Mass spectrometers of high transmission and high resolving power. Nucl Instr Meth 187:127–136

Matsuda H, Naito M, Takeuchi M (1974) Advanced virtual image double focussing mass spectrometer. Adv Mass Spectrom 6:407–412

Matz G, Schröder W (1996) Fast GC/MS field screening for excavation and bioredmediation of contaminated soil. Field Anal Chem Technol 1:77–85

Matz G, Schröder W (1997) Fast detection of wood preservatives on waste wood with GC/MS, GC/ECD and ion mobility spectrometry. Conference "Field analytical methods for hazardous wastes and toxic chemicals", Las Vegas

Matz LM, Hill HH (2001) Evaluation of opiate separation by high-resolution electrospray ionization-ion mobility spectrometry / mass spectrometry. Anal Chem 73:1664–1669

Mayor U, Johnson CM, Daggett V, Fersht AR (2000) Protein folding and unfolding in microseconds to nanoseconds by experiment and simulation. Proc Natl Acad Sci USA 97:13518–13522

McCraith S, Holtzman T, Moss B, Fields S (2000) Genome-wide analysis of vaccinia virus protein-protein interactions. Proc Natl Acad Sci USA 97:4879–4884

McPherson A, Malkin AJ, Kuznetsov YG (2000) Atomic force microscopy in the study of macromolecular crystal growth. Annu Rev Biophys Biomol Struct 29:361–410

McPherson A, Malkin AJ, Kuznetsov YG, Plomp M (2001) Atomic force microscopy applications in macromolecular crystallography. Acta Crystallogr D Biol Crystallogr 57:1053–1060

Megerle CA, Cohn DB (2000) Ion mobility sensors and spectrometers having a corona discharge ionization source. US Patent 6,100,698

Meixner AJ, Kneppe H (1998) Scanning near-field optical microscopy in cell biology and microbiology. Cell Mol Biol 44:673–688

Meland BC (1980) Apparatus for electrophysiological stimulation. US Patent 4,227,516

Merkel R, Nassoy P, Leung A, Ritchie K, Evans E (1999) Energy landscapes of receptor-ligand bonds explored with dynamic force spectroscopy. Nature 397:50–53

Mezzetti A, Nabedryk E, Breton J, Okamura MY, Paddock ML, Giacometti G, Leibl W (2002) Rapid-scan Fourier transform infrared spectroscopy shows coupling of Glu-L212 protonation and electron transfer to Q(B) in *Rhodobacter sphaeroides* reaction centers. Biochim Biophys Acta 1553:320–330

Michener CM, Ardekani AM, Petricoin EF 3rd, Liotta LA, Kohn EC (2002) Genomics and proteomics: application of novel technology to early detection and prevention of cancer. Cancer Detect Prev 26:249–255

Miki A, Keller T, Regenscheit P, Dirnhofer R, Tatsuno M, Katagi M, Nishikawa M, Tsuchihashi H (1997) Application of ion mobility spectrometry to the rapid screening of methamphetamine incorporated in hair. J Chromatogr B 692:319–328

Miki A, Tatsuno M, Katagie M, Nishikawa M, Tsuchihashi H (1998) Analysis of illicit drugs by ion mobility spectrometry. J Toxicol - Toxin Rev 17:93–93

Miller LD, Putthanarat S, Eby RK, Adams WW (1999) Investigation of the nanofibrillar morphology in silk fibers by small angle X-ray scattering and atomic force microscopy. Int J Biol Macromol 24:159–165

Mills G, Zhou H, Midha A, Donaldson L, Weaver JMR (1998) Scanning thermal microscopy using batch fabricated thermocouple probes. Appl Phys Lett 72:2900–2902

Mills G, Weaver JMR, Harris G, Chen W, Carrejo J, Johnson L, Rogers B (1999) Detection of subsurface voids using scanning thermal microscopy. Ultramicroscopy 80:7–11

Mirny L, Shakhnovich E (2001) Protein folding theory: from lattice to all-atom models. Annu Rev Biophys Biomol Struct 30:361–396

Mitsuoka Y, Niwa T, Ichihara S, Kato K, Muramatsu H, Nakajima K, Shikida M, Sato K (2001) Microfabricated silicon dioxide cantilever with subwavelength aperture. J Microsc 202:12–15

Mo W, Karger BL (2002) Analytical aspects of mass spectrometry and proteomics. Curr Opin Chem Biol 6:666–675

Mollica V, Borassi A, Relini A, Cavalleri O, Bolognesi M, Rolandi R, Gliozzi A (2001)

An atomic force microscopy investigation of protein crystal surface topography. Eur Biophys J 30:313–318

Moritz R, Reinstadler D, Fabian H, Naumann D (2002) Time-resolved FTIR difference spectroscopy as tool for investigating refolding reactions of ribonuclease T1 synchronized with *trans* → *cis* prolyl isomerization. Biopolymers 67:145–155

Morris VJ, Kirby AR, Gunning AP (1999) Using atomic force microscopy to probe food biopolymer functionality. Scanning 21:287–292

Morrison RS, Kinoshita Y, Johnson MD, Uo T, Ho JT, McBee JK, Conrads TP, Veenstra TD (2002) Proteomic analysis in the neurosciences. Mol Cell Proteomics 1:553–560

Morton CJ, Pugh DJR, Brown ELJ, Kahmann JD, Renzoni DAC, Campbell ID (1996) Solution structure and peptide binding of the SH3 domain from human Fyn. Structure 4:705–714

Mui C, Han JH, Wang GT, Musgrave CB, Bent SF (2002) Proton transfer reactions on semiconductor surfaces. J Am Chem Soc 124:4027–4038

Muñoz V, Eaton WA (1999) A simple model for calculating the kinetics of protein folding from three-dimensional structures. Proc Natl Acad Sci USA 96:11311–11316

Muramatsu H, Homma K, Chiba N, Yamamoto N, Egawa A (1999) Dynamic etching method for fabricating a variety of tip shapes in the optical fiber probe of a scanning near-field optical microscope. J Microsc 194:383–387

Muroga Y (2001) Derivation of the small angle X-ray scattering functions for local conformations of polypeptide chains in solution. Biopolymers 59:320–329

Natsume T, Yamauchi Y, Nakayama H, Shinkawa T, Yanagida M, Takahashi N, Isobe T (2002) A direct nanoflow liquid chromatography-tandem mass spectrometry system for interaction proteomics. Anal Chem 74:4725–4733

Nemeth-Cawley JF, Rouse JC (2002) Identification and sequencing analysis of intact proteins via collision-induced dissociation and quadrupole time-of-flight mass spectrometry. J Mass Spectrom 37:270–282

Niggemann M, Steipl B (2000) Exploring local and nonlocal interactions for protein stability by structural motif engineering. J Mol Biol 296:181–195

Nilges M, Macias MJ, O'Donoghue SI, Oschkinat H (1997) Automated NOESY interpretation with ambiguous distance restraints: the refined NMR solution structure of the pleckstrin homology domain from beta-spectrin. J Mol Biol 269:408–422

Nilsson CL (2002) Bacterial proteomics and vaccine development. Am J Pharmacogenomics 2:59–65

Nilsson T, Bassani MR, Larsen TO, Montanarella L (1996) Classification of species in the genus *Penicillium* by Curie point pyrolysis / mass spectrometry followed by multivariate analysis and artificial neural networks. J Mass Spectrom 31:1422–1428

Noinville S, Revault M, Baron MH (2002) Conformational changes of enzymes adsorbed at liquid-solid interface: relevance to enzymatic activity. Biopolymers 67:323–326

Nölting B (1991) Development of a novel spectrometer for the simultaneous measurement of absorption and circular dichroism (in German). PhD thesis, University of Bochum.

Nölting B (1998) Structural resolution of the folding pathway of a protein by correlation of Φ-values with inter-residue contacts. J Theor Biol 194:419–428

Nölting B (1999a) Analysis of the folding pathway of chymotrypsin inhibitor by correlation of Φ-values with inter-residue contacts. J Theor Biol 197:113–121

Nölting B (1999b) Protein Folding Kinetics: Biophysical Methods. Springer, Berlin Heidelberg New York

Nölting B (2002) Thought transmission (in German). Patent application

Nölting B, Andert K (2000) Mechanism of protein folding. Proteins 41:288–298

Nölting B, Golbik R, Fersht AR (1995) Submillisecond events in protein folding.
Proc Natl Acad Sci USA 92:10668–10672

Nölting B, Golbik R, Neira JL, Soler-Gonzalez AS, Schreiber G, Fersht AR (1997a) The
folding pathway of a protein at high resolution from microseconds to seconds.
Proc Natl Acad Sci USA 94:826–830

Nölting B, Golbik R, Soler-González AS, Fersht AR (1997b) Circular dichroism of
denatured barstar shows residual structure. Biochemistry 36:9899–9905

Nölting B, Schälike W, Hampel P, Grundig F, Gantert S, Sips N, Bandlow W, Qi PX
(2003) Structural determinants of the rate of protein folding. J Theor Biol, in press

Norledge B, Mayr EM, Glockshuber R, Bateman OA, Slingsby C, Jaenicke R,
Driessen HPC (1996) The X-ray structures of two mutant crystallin domains shed light
on the evolution of multi-domain proteins. Nature Struct Biol 3:267–274

Nyman TA (2001) The role of mass spectrometry in proteome studies. Biomol Eng
18:221–227

Odom TW, Huang JL, Lieber CM (2002) Single-walled carbon nanotubes: from
fundamental studies to new device concepts. Ann N Y Acad Sci 960:203–215

Oesterhelt F, Oesterhelt D, Pfeiffer M, Engel A, Gaub HE, Müller DJ (2000) Unfolding
pathways of individual bacteriorhodopsins. Science 288:143–146

Ogden ID, Strachan NJC (1993) Enumeration of *Escherichia coli* in cooked and raw meats
by ion mobility spectrometry. J Applied Bacteriology 74:402–405

Orengo CA, Jones DT, Thornton JM (1994) Protein superfamilies and domain superfolds.
Nature 372:631–634

Oroudjev E, Soares J, Arcdiacono S, Thompson JB, Fossey SA, Hansma HG (2002)
Segmented nanofibers of spider dragline silk: atomic force microscopy and single-
molecule force spectroscopy. Proc Natl Acad Sci USA 99:6460–6465

Oubridge C, Ito N, Evans PR, Teo CH, Nagai K (1994) Crystal structure at 1.92 Å
resolution of the RNA-binding domain of the U1A spliceosomal protein complexed
with an RNA hairpin. Nature 372:432–438

Panick G, Malessa R, Winter R, Rapp G, Frye KJ, Royer CA (1998) Structural
characterization of the pressure-denatured state and unfolding/refolding kinetics of
staphylococcal nuclease by synchrotron small angle X-ray scattering and Fourier
transform infrared spectroscopy. J Mol Biol 275:389–402

Panick G, Malessa R, Winter R (1999a) Differences between the pressure- and
temperature-induced denaturation and aggregation of β-lactoglobulin A, B, and AB
monitored by FTIR spectroscopy and small angle X-ray scattering. Biochemistry
38:6512–6519

Panick G, Vidugiris GJ, Malessa R, Rapp G, Winter R, Royer CA (1999b) Exploring the
temperature-pressure phase diagram of staphylococcal nuclease. Biochemistry
38:4157–4164

Park SJ, Taton TA, Mirkin CA (2002) Array-based electrical detection of DNA with
nanoparticle probes. Science 295:1503–1506

Pastore A, Saudek V, Ramponi G, Williams RJP (1992) Three-dimensional structure of
acylphosphatase refinement and structure analysis. J Mol Biol 224:427–440

Pereira RS (2001) Atomic force microscopy as a novel pharmacological tool. Biochem
Pharmacol 62:975–983

Perez J, Defrenne S, Witz J, Vachette P (2000) Detection and characterization of an intermediate conformation during the divalent ion-dependent swelling of tomato bushy stunt virus. Cell Mol Biol 46:937–948

Perkins WD (1986) Fourier transform infrared spectroscopy. J Chem Education 63:A5–A10

Perkins G, Renken C, Martone ME, Young SJ, Ellisman M, Frey TG (1997a) Electron tomography of neuronal mitochondria: three-dimensional structure and organization of cristae and membrane contacts. J Struct Biol 119:260–272

Perkins G, Renken CW, Song JY, Frey TG, Young SJ, Lamont S, Martone ME, Lindsey S, Ellisman MH (1997b) Electron tomography of large, multicomponent biological structures. J Struct Biol 120:219–227

Perutz MF, Rossmann MG, Cullis AF, Muirhead G, Will G, North AT (1960) Structure of haemoglobin: a three-dimensional Fourier synthesis at 5.5 Å resolution, obtained by X-ray analysis. Nature 185:416–422

Petsko GA, Ringe D (2000) Observation of unstable species in enzyme-catalyzed transformations using protein crystallography. Curr Opin Chem Biol 4:89–94

Philippsen A, Im W, Engel A, Schirmer T, Roux B, Müller DJ (2002) Imaging the electrostatic potential of transmembrane channels: atomic probe microscopy of OmpF porin. Biophys J 82:1667–1676

Phillips J, Gormally J (1992) The laser desorption of organic molecules in ion mobility spectrometry. Int J Mass Spectrom Ion Processes 112:205–214

Phillips GN Jr, Arduini RM, Springer BA, Sligar SG (1990) Crystal structure of myoglobin from a synthetic gene. Proteins 7:358–365

Phizicky EM, Martzen MR, McCraith SM, Spinelli SL, Xing F, Shull NP, Van Slyke C, Montagne RK, Torres FM, Fields S, Grayhack EJ (2002) Biochemical genomics approach to map activities to genes. Methods Enzymol 350:546–559

Pillutla RC, Goldstein NI, Blume AJ, Fisher PB (2002) Target validation and drug discovery using genomic and protein-protein interaction technologies. Expert Opin Ther Targets 6:517–531

Plaxco KW, Simons KT, Baker D (1998) Contact order, transition state placement and the refolding rates of single-domain proteins. J Mol Biol 277:985–994

Pohl DW, Denk W, Lanz M (1984) Optical stethoscopy: image recording with resolution λ/20. Appl Phys Lett 44:651–653

Powell KD, Wales TE, Fitzgerald MC (2002) Thermodynamic stability measurements on multimeric proteins using a new H/D exchange- and matrix-assisted laser desorption/ionization (MALDI) mass spectrometry-based method. Protein Sci 11:841–851

Pralle A, Florin EL (2002) Cellular membranes studied by photonic force microscopy. Methods Cell Biol 68:193–212

Prechtel K, Bausch AR, Marchi-Artzner V, Kantlehner M, Kessler H, Merkel R (2002) Dynamic force spectroscopy to probe adhesion strength of living cells. Phys Rev Lett 89:028101-1–4

Preston R (1998) The bioweaponeers. The New Yorker, March 9:52–65

Prokop A, Holland CA, Kozlov E, Moore B, Tanner RD (2001) Water-based nanoparticulate polymeric system for protein delivery. Biotechnol Bioeng 75:228–232

Prusiner SB (ed) (1999) Prion Biology and Diseases. Cold Spring Harbor Monograph Series, No 38.

Purves RW, Barnett DA, Ells B, Guevremont R (2000) Investigation of bovine ubiquitin conformers separated by high-field asymmetric waveform ion mobility spectrometry:

cross section measurements using energy-loss experiments with a triple quadrupole mass spectrometer. J Am Soc Mass Spectrom 11:738–745

Ranson NA, White HE, Saibil HR (1998) Chaperonins. Biochem J 333:233–242

Razatos A (2001) Application of atomic force microscopy to study initial events of bacterial adhesion. Methods Enzymol 337:276–285

Reimers JR, Shapley WA, Lambropoulos N, Hush NS (2002) An atomistic approach to conduction between nanoelectrodes through a single molecule. Ann N Y Acad Sci 960:100–130

Renfrey S, Featherstone J (2002) Structural proteomics. Nat Rev Drug Discov 1:175–176

Richards FM (1974) The interpretation of protein structures: Total volume, group volume distributions and packing density. J Mol Biol 82:1–14

Riddle DS, Grantcharova VP, Santiago JV, Alm E, Ruczinski I, Baker D (1999) Experiment and theory highlight role of native state topology in SH3 folding. Nature Struct Biol 6:1016–1024

Riedel M, Müller B, Wintermantel E (2001) Protein adsorption and monocyte activation on germanium nanopyramids. Biomaterials 22:2307–2316

Rief M, Gautel M, Oesterhelt F, Fernandez JM, Gaub HE (1997) Reversible unfolding of individual titin immunoglobulin domains by AFM. Science 276:1109–1112

Riek R, Hornemann S, Wider G, Billeter M, Glockshuber R, Wüthrich K (1996) NMR structure of the mouse prion protein domain PrP(121–321). Nature 382:180–182

Riek R, Wider G, Billeter M, Hornemann S, Glockshuber R, Wüthrich K (1998) Prion protein NMR structure and familial human spongiform encephalopathies. Proc Natl Acad Sci USA 95:11667–11672

Riekel C, Vollrath F (2001) Spider silk fibre extrusion: combined wide- and small angle X-ray microdiffraction experiments. Int J Biol Macromol 29:203–210

Rinaldi R, Branca E, Cingolani R, Di Felice R, Calzolari A, Molinari E, Masiero S, Spada G, Gottarelli G, Garbesi A (2002) Biomolecular electronic devices based on self-organized deoxyguanosine nanocrystals. Ann N Y Acad Sci 960:184–192

Rinia HA, Boots JW, Rijkers DT, Kik RA, Snel MM, Demel RA, Killian JA, van der Eerden JP, de Kruijff B (2002) Domain formation in phosphatidylcholine bilayers containing transmembrane peptides: specific effects of flanking residues. Biochemistry 41:2814–2824

Rinnerthaler S, Roschger P, Jakob HF, Nader A, Klaushofer K, Fratzl P (1999) Scanning small angle X-ray scattering analysis of human bone sections. Calcif Tissue Int 64:422–429

Riske KA, Amaral LQ, Lamy-Freund MT (2001) Thermal transitions of DMPG bilayers in aqueous solution: SAXS structural studies. Biochim Biophys Acta 1511:297–308

Rohlff C, Southan C (2002) Proteomic approaches to central nervous system disorders. Curr Opin Mol Ther 4:251–258

Roseman A, Chen S, White H, Braig K, Saibil HR (1996) The chaperonin ATPase cycle: mechanism of allosteric switching and movements of substrate-binding domains in GroEL. Cell 87:241–251

Russell R, Millett IS, Doniach S, Herschlag D (2000) Small angle X-ray scattering reveals a compact intermediate in RNA folding. Nature Struct Biol 7:367–370

Rye HS, Roseman AM, Chen S, Furtak K, Fenton WA, Saibil HR, Horwich AL (1999) GroEL-GroES cycling: ATP and nonnative polypeptide direct alternation of folding-active rings. Cell 97:325–338

Saibil H (2000a) Molecular chaperones: containers and surfaces for folding, stabilizing or unfolding proteins. Curr Opin Struct Biol 10:251–258

Saibil HR (2000b) Conformational changes studied by cryo-electron microscopy. Nature Struct Biol 7:711–714

Sandhu KK, McIntosh CM, Simard JM, Smith S, Rotello VM (2002) Gold nanoparticle-mediated transfection of mammalian cells. Bioconjug Chem 13:3–6

Sanger F (1988) Sequences, sequences, and sequences. Annu Rev Biochem 57:1–28

Sanger F, Nicklen S, Coulson AR (1977) DNA sequencing with chain-terminating inhibitors. Proc Natl Acad Sci USA 74:5463–5467

Sano Y, Inoue H, Hiragi Y (1999) Differences of reconstitution process between tobacco mosaic virus and cucumber green mottle mosaic virus by synchrotron small angle X-ray scattering using low-temperature quenching. J Protein Chem 18:801–805

Sato M, Hida M, Nagase H (2001) Analysis of pyrolysis products of dimethylamphetamine. J Anal Toxicol 25:304–309

Saurina J, Hernandez-Cassou S (1999) Flow-injection and stopped-flow completely continuous flow spectrometric determination of aniline and cyclohexylamine. Anal Chim Acta 396:151–159

Scheuring S, Stahlberg H, Chami M, Houssin C, Rigaud JL, Engel A (2002) Charting and unzipping the surface layer of *Corynebacterium glutamicum* with the atomic force microscope. Mol Microbiol 44:675–684

Schindelin H, Marahiel MA, Heinemann U (1993) Universal nucleic acid-binding domain revealed by crystal structure of the *B. subtilis* major cold-shock protein. Nature 364:164–168

Schlichting I, Berendzen J, Chu K, Stock AM, Maves SA, Benson DE, Sweet RM, Ringe D, Petsko GA, Sligar SG (2000) The catalytic pathway of cytochrome P450$_{cam}$ at atomic resolution. Science 287:1615–1622

Schmid MB (2002) Structural proteomics: the potential of high-throughput structure determination. Trends Microbiol. 10(Suppl):S27–31

Schmitke JL, Stern LJ, Klibanov AM (1997) The crystal structure of subtilisin Carlsberg in anhydrous dioxane and its comparison with those in water and acetonitrile. Proc Natl Acad Sci USA 94:4250–4255

Schmitke JL, Stern LJ, Klibanov AM (1998) Comparison of X-ray crystal structures of an acyl-enzyme intermediate of subtilisin Carlsberg formed in anhydrous acetonitrile and in water. Proc Natl Acad Sci USA 95:12918–12923

Schnurpfeil R, Klepel S (2000) Radioactivity ion sources for miniaturized ion mobility spectrometers. US Patent 6,064,070

Schön JH, Meng H, Bao Z (2001a) Self-assembled monolayer organic field-effect transistors. Nature 413:713–716

Schön JH, Meng H, Bao Z (2001b) Self-assembled monolayer organic field-effect transistors. Nature 414:470–470

Schönbrunn E, Svergun DI, Amrhein N, Koch MH (1998) Studies on the conformational changes in the bacterial cell wall biosynthetic enzyme UDP-N-acetylglucosamine enolpyruvyltransferase (MurA). Eur J Biochem 15:406–412

Schröder W, Matz G, Kubler J (1998) Fast detection of preservatives on waste wood with GC/MS, GC-ECD and ion-mobility spectrometry. Field Anal Chem Technol 2:287–297

Schurmann G, Noell W, Staufer U, de Rooij NF (2000) Microfabrication of a combined AFM-SNOM sensor. Ultramicroscopy 82:33–38

Schwartz DE, Mancinelli RL, White MR (1995) Search for life on Mars: evaluation of techniques. Adv Space Res 15:193–197

Schweitzer-Stenner R (2002) Dihedral angles of tripeptides in solution directly determined by polarized Raman and FTIR spectroscopy. Biophys J 83:523–532

Schwesinger F, Ros R, Strunz T, Anselmetti D, Güntherodt HJ, Honegger A, Jermutus L, Tiefenauer L, Plückthun A (2000) Unbinding forces of single antibody-antigen complexes correlate with their thermal dissociation rates. Proc Natl Acad Sci USA 97:9972–9977

Scott DJ, Grossmann JG, Tame JR, Byron O, Wilson KS, Otto BR (2002) Low resolution solution structure of the Apo form of *Escherichia coli* haemoglobin protease Hbp. J Mol Biol 315:1179–1187

Seeman NC, Belcher AM (2002) Emulating biology: building nanostructures from the bottom up. Proc Natl Acad Sci USA 99:6451–6455

Segel DJ, Eliezer D, Uversky V, Fink AL, Hodgson KO, Doniach S (1999) Transient dimer in the refolding kinetics of cytochrome c characterized by small angle X-ray scattering. Biochemistry 38:15352–15359

Sekatskii SK, Dietler G (1999) Near-field optical excitation as a dipole-dipole energy transfer process. J Microsc 194:255–259

Seong GH, Kobatake E, Miura K, Nakazawa A, Aizawa M (2002) Direct atomic force microscopy visualization of integration host factor-induced DNA bending structure of the promoter regulatory region on the *Pseudomonas* TOL plasmid. Biochem Biophys Res Commun 291:361–366

Service RF (2001) Breakthrough of the year. Molecules get wired. Science 294:2442–2443

Service RF (2002) Analytical chemisty – new test could speed bioweapon detection. Science 295:1447–1447

Shakhnovich EI (1997) Theoretical studies of protein-folding thermodynamics and kinetics. Curr Opin Struct Biol 7:29–40

Shakhnovich E, Abkevich V, Ptitsyn O (1996) Conserved residues and the mechanism of protein folding. Nature 379:96–98

Shevchenko A, Chernushevic I, Shevchenko A, Wilm M, Mann M (2002) "De novo" sequencing of peptides recovered from in-gel digested proteins by nanoelectrospray tandem mass spectrometry. Mol Biotechnol 20:107–118

Shilton B, Svergun DI, Volkov VV, Koch MH, Cusack S, Economou A (1998) *Escherichia coli* SecA shape and dimensions. FEBS Lett 436:277–282

Shimonishi Y, Hong YM, Kitagishi T, Matsuo T, Matsuda H, Katakuse I (1980) Sequencing of peptide mixtures by Edman degradation and field-desorption mass spectrometry. Eur J Biochem 112:251–264

Shumate CB, Hill HH (1989) Coronaspray nebulization and ionization of liquid samples for ion mobility spectrometry. Anal Chem 61:601–606

Simpson AA, Tao YZ, Leiman PG, Badasso MO, He Y, Jardine PJ, Olson NH, Morais MC, Grimes S, Anderson DL, Baker TS, Rossmann MG (2000) Structure of the bacteriophage Φ29 DNA packaging motor. Nature 408:745–750

Sisson PR, Freeman R, Magee JG, Lightfoot NF (1991) Differentiation between mycobacteria of the *Mycobacterium tuberculosis* complex by pyrolysis mass spectrometry. Tubercle 72:206–209

Smith DE, Tans SJ, Smith SB, Grimes S, Anderson DL, Bustamante C (2001) The bacteriophage straight Φ29 portal motor can package DNA against a large internal force. Nature 413:748–752

Smith SB, Cui Y, Bustamante C (1996) Overstreaching B-DNA: the elastic resonse of individual double-stranded and single-stranded DNA molecules. Science 271:795–799

Smith GB, Eiceman GA, Walsh MK, Critz SA, Andazola E, Ortega E, Cadena F (1997) Detection of *Salmonella typhimurium* by hand-held ion mobility spectrometer: a quantitative assessment of response characteristics. Field Anal Chem Technol 1:213–226

Snabe T, Petersen SB (2002) Application of infrared spectroscopy (attenuated total reflection) for monitoring enzymatic activity on substrate films. J Biotechnol 95:145–155

Snow CD, Nguyen H, Pande VS, Gruebele M (2002) Absolute comparison of simulated and experimental protein-folding dynamics. Nature 420:102–106

Snyder AP, McClennen WH, Dworzanski JP, Meuzelaar HL (1990) Characterization of underivatized lipid biomarkers from microorganisms with pyrolysis short-column gas chromatography / ion trap mass spectrometry. Anal Chem 62:2565–2573

Snyder AP, Miller M, Shoff DB, Eiceman GA, Blyth DA, Parsons JA (1991a) Enzyme-substrate assay for the qualitative detection of microorganisms by ion mobility spectrometry. J Microbiol Methods 14:21–32

Snyder AP, Shoff DB, Eiceman GA, Blyth DA, Parsons JA (1991b) Detection of bacteria by ion mobility spectrometry. Anal Chem 63:526–529

Snyder AP, Harden CS, Brittain AH, Kim MG, Arnold NS, Meuzelaar HLC (1993) Portable hand-held gas chromatography / ion mobility spectrometry device. Anal Chem 65:299–306

Snyder AP, Blyth DA, Parsons JA (1996a) Ion mobility spctrometry as an immunoassay detection technique. J Microbiol Methods 27:81–88

Snyder AP, Thornton SN, Dworzanski JP, Meuzelaar HLC (1996b) Detection of picolinic acid biomarker in *Bacillus* spores using a potentially field-portable pyrolysis gas chromatography / ion mobility spectrometer. Field Anal Chem Technol 1:49–59

Snyder AP, Maswadeh WM, Parson JP, Tripathi A, Meuzelaar HLC, Dworzanski J, Kim MG (1999) Field detection of *Bacillus* spore aerosols with stand-alone pyrolysis gas chromatography / ion mobility spectrometry. Field Anal Chem Technol 3:315–326

Snyder AP, Maswadeh WM, Tripathi A, Dworzanski JP (2000) Detection of gram-negative *Erwinia herbicola* outdoor aerosols with pyrolysis gas chromatography / ion mobility spectrometry. Field Anal Chem Technol 4:111–126

Snyder AP, Tripathi A, Maswadeh WM, Ho J, Spence M (2001) Field detection and identification of a bioaerosol suite by pyrolysis / gas chromatography / ion mobility spectrometry. Field Anal Chem Technol 5:190–204

Spangler GE (1982) Membrane interface for ion mobility detector cells. US Patent 4,311,669

Spangler GE (1992a) Preconcentrator for ion mobility spectrometer. US Patent 5,083,019

Spangler GE (1992b) Space charge effects in ion mobility spectrometry. Anal Chem 64:1312–1312

Spangler GE, Carrico JP (1983) Membrane inlet for ion mobility spectrometry (plasma chromatography). Int J Mass Spectrom Ion Phys 52:267–287

Spangler GE, Roehl JE, Patel GB, Dorman A (1994) Photoionization ion mobility spectrometer. US Patent 5,338,931

Srajer V, Teng TY, Ursby T, Pradervand C, Ren Z, Adachi SI, Schildkamp W, Bourgeois D, Wulff M, Moffat K (1996) Photolysis of the carbon monoxide complex of myoglobin: nanosecond time-resolved crystallography. Science 274:1726–1729

Stachelberger H (2001) personal communication.

Stagljar I, Fields S (2002) Analysis of membrane protein interactions using yeast-based technologies. Trends Biochem Sci 27:559–63

Steinfeld JI, Wormhoudt J (1998) Explosives detection: a challenge for physical chemistry. Annu Rev Phys Chem 49:203–232

Stephenson JL, McLuckey SA, Reid GE, Wells JM, Bundy JL (2002) Ion/ion chemistry as a top-down approach for protein analysis. Curr Opin Biotechnol 13:57–64

St Louis RH, Hill HH (1990) Ion mobility spectrometry in analytical chemistry. CRC Crit Rev Anal Chem 21:321–355

Stöckle RM, Fokas C, Deckert V, Zenobi R, Sick B, Hecht B, Wild UP (1999a) High quality near-field optical probes by tube etching. Appl Phys Lett 75:160–162

Stöckle RM, Schaller N, Deckert V, Fokas C, Zenobi R (1999b) Brighter near-field optical probes by means of improving the optical destruction threshold. J Microsc 194:378–382

Stocklin PL (1989) Hearing device. US Patent 4,858,612

Stoeva S, Idakieva K, Betzel C, Genov N, Voelter W (2002) Amino acid sequence and glycosylation of functional unit RtH2-e from *Rapana thomasiana* (gastropod) hemocyanin. Arch Biochem Biophys 399:149–158

Stone E, Gillig KJ, Ruotolo B, Fuhrer K, Gonin M, Schultz A, Russell DH (2001) Surface-induced dissociation on a MALDI ion mobility / orthogonal time-of-flight mass spectrometer: sequencing peptides from an "in-solution" protein digest. Anal Chem 73:2233–2238

Strachan NJC, Nicholson FJ, Ogden ID (1995) An automated sampling system using ion mobility spectrometry for the rapid detection of bacteria. Anal Chim Acta 313:63–67

Strunz T, Oroszlan K, Schäfer R, Güntherodt HJ (1999) Dynamic force spectroscopy of single DNA molecules. Proc Natl Acad Sci USA 96:11277–11282

Sultana S, Magee JT, Duerden B (1995) Analysis of Bacteroides species by pyrolysis mass spectrometry. Clin Infect Dis 20 Suppl 2:S122–S127

Swedberg SA, Kaltenbach P, Witt KE, Bek F, Mittelstadt LS (1996) Fully integrated iniaturized planar liquid sample handling and analysis device. US Patent 5,571,410

Swedberg SA, Brennen RA (2001) Device for high throughput sample processing, analysis and collection, and methods of use thereof. US Patent 6,240,790

Synge EH (1928) A suggested method for extending microscopic resolution into the ultra-microscopic region. Phil Mag 6:356–362

Talapatra A, Rouse R, Hardiman G (2002) Protein microarrays: challenges and promises. Pharmacogenomics 3:527–536

Tanaka S, Scheraga HA (1975) Model of protein folding: Inclusion of short-, medium-, and long-range interactions. Proc Natl Acad Sci USA 72:3802–3806

Tanaka S, Scheraga HA (1977) Hypothesis about the mechanism of protein folding. Macromolecules 10:291–304

Taylor J, Goodacre R, Wade WG, Rowland JJ, Kell DB (1998) The deconvolution of pyrolysis mass spectra using genetic programming: application to the identification of some Eubacterium species. FEMS Microbiol Lett 160:237–246

Taylor SJ (1996) Introduction of samples into an ion mobility spectrometer. US Patent 5,574,277

Taylor SJ, Turner RB (1999) Ion mobility spectrometers. US Patent 5,952,652

Tcherkasskaya O, Uversky VN (2001) Denatured collapsed states in protein folding: example of apomyoglobin. Proteins 44:244–254

Tilleman K, Van den Haute C, Geerts H, van Leuven F, Esmans EL, Moens L (2002) Proteomics analysis of the neurodegeneration in the brain of tau transgenic mice. Proteomics 2:656–665

Timmins EM, Goodacre R (1997) Rapid quantitative analysis of binary mixtures of *Escherichia coli* strains using pyrolysis mass spectrometry with multivariate calibration and artificial neural networks. J Appl Microbiol 83:208–218

Tiner WJS, Potaman VN, Sinden RR, Lyubchenko YL (2001) The structure of intramolecular triplex DNA: atomic force microscopy study. J Mol Biol 314:353–357

Toledo-Crow R, Yang PC, Chen Y, Vaez-Iravani M (1992) Near-field differential scanning optical microscope with atomic force regulation. Appl Phys Lett 60:2957–2959

Torres J, Briggs JA, Arkin IT (2002) Multiple site-specific infrared dichroism of CD3-zeta, a transmembrane helix bundle. J Mol Biol 316:365–374

Tripathi A, Maswadeh WM, Snyder AP (2001) Optimization of quartz tube pyrolysis atmospheric pressure ionization mass spectrometry for the generation of bacterial biomarkers. Rapid Commun Mass Spectrom 15:1672–1680

Trudel E, Gallant J, Mons S, Mioskowski C, Lebeau L, Jeuris K, Foubert P, De Schryver F, Salesse C (2001) Design of functionalized lipids and evidence for their binding to photosystem II core complex by oxygen evolution measurements, atomic force microscopy, and scanning near-field optical microscopy. Biophys J 81:563–571

Turner DR (1983) Etch procedure for optical fibers. US Patent 4,469,554

Turner BR (1993) Ion mobility detector. US Patent 5,227,628

Unger R, Moult J (1996) Local interactions dominate folding in a simple protein model. J Mol Biol 259:988–994

Valle F, Dietler G, Londei P (2001) Single-molecule imaging by atomic force microscopy of the native chaperonin complex of the thermophilic archaeon *Sulfolobus solfataricus*. Biochem. Biophys Res Commun 288:258–262

van den Berg B, Wain R, Dobson CM, Ellis RJ (2000) Macromolecular crowding perturbs protein refolding kinetics: implications for folding inside the cell. EMBO J 19:3870–3875

van Duyne GD, Standaert RF, Karplus PA, Schreiber SL, Clardy J (1991) Atomic structure of FKBP-FK506, an immunophilin-immunosuppressant complex. Science 252:839–842

van Kempen TA, Powers WJ, Sutton AL (2002) Technical note: Fourier transform infrared (FTIR) spectroscopy as an optical nose for predicting odor sensation. J Anim Sci 80:1524–1527

van Nuland NAJ, Hangyi IW, van Schaik RC, Berendsen HJC, van Gunsteren WF, Scheek RM, Robillard GT (1994) The high-resolution structure of the histidine-containing phosphocarrier protein HPr from *Escherichia coli* determined by restrained molecular dynamics from nuclear magnetic resonance nuclear Overhauser effect data. J Mol Biol 237:544–559

Veerman JA, Otter AM, Kuipers L, van Hulst NF (1998) High definition aperture probes for near-field optical microscopy fabricated by focused ion beam milling. Appl Phys Lett 72:3115–3117

Vendrell J, Billeter M, Wider G, Aviles FX, Wüthrich K (1991) The NMR structure of the activation domain isolated from porcine procarboxypeptidase B. EMBO J 10:11–15

Vijay-Kumar S, Bugg CE, Cook WJ (1987) Structure of ubiquitin refined at 1.8 Å resolution. J Mol Biol 194:531–544

Voigt J, Schrötter T (1999) Phonon assisted exciton transitions on LHC-II complexes – a long wavelength absorption mechanism by cooperative action of photons and protein vibrations. Zeitschrift Phys Chem 211:181–191

Volz K, Matsumura P (1991) Crystal structure of *Escherichia coli* CheY refined at 1.7 Å resolution. J Biol Chem 266:15511–15519

von Ardenne M (1940) Elektronen-Übermikroskopie, Springer, Berlin

Vora KN, Carrico JP Sr, Spangler GE, Campbell DN, Martin CE (1987) Ion mobility spectrometer. US Patent 4,712,008

Vytvytska O, Nagy E, Bluggel M, Meyer HE, Kurzbauer R, Huber LA, Klade CS (2002) Identification of vaccine candidate antigens of Staphylococcus aureus by serological proteome analysis. Proteomics 2:580–590

White HE, Chen S, Roseman AM, Yifrach O, Horovitz A, Saibil H (1997) Structural basis of allosteric changes in the GroEL mutant Arg197→Ala. Nature Struct Biol 4:690–694

Whitelegge JP, le Coutre J (2002) Proteomics. Making sense of genomic information for drug discovery. Am J Pharmacogenomics 1:29–35

Wikström M, Sjöbring U, Drakenberg T, Forsén S, Björck L (1995) Mapping of the immunoglobulin light chain-binding site of protein L. J Mol Biol 250:128–133

Williams S, Causgrove TP, Gilmanshin R, Fang KS, Callender RH, Woodruff WH, Dyer RB (1996) Fast events in protein folding: helix melting and formation in a small peptide. Biochemistry 35:691–697

Williams JC, Zeelen JP, Neubauer G, Vriend G, Backmann J, Michels PA, Lambeir AM, Wierenga RK (1999) Structural and mutagenesis studies of leishmania triosephosphate isomerase: a point mutation can convert a mesophilic enzyme into a superstable enzyme without losing catalytic power. Protein Eng 12:243–250

Williams BH, Hathout Y, Fenselau C (2002) Structural characterization of lipopeptide biomarkers isolated from *Bacillus globigii*. J Mass Spectrom 37:259–264

Williamson RL, Miles MJ (1996) Melt-drawn scanning near-field optical microscopy probe profiles. J Appl Phys 80:4804–4812

Williamson RL, Brereton LJ, Antognozzi M, Miles MJ (1998) Are artefacts in scanning near-field optical microscopy related to the misuse of shear force? Ultramicroscopy 71:165–175

Wilmot CM, Pearson AR (2002) Cryocrystallography of metalloprotein reaction intermediates. Curr Opin Chem Biol 6:202–207

Wolynes PG, Luthey-Schulten Z, Onuchic JN (1996) Fast folding experiments and the topography of protein-folding energy landscapes. Chem Biol 3:425–432

Wu C, Hill HH, Gamerdinger AP (1998a) Electrospay ionization / ion mobility spectrometry as a field monitoring method for the detection of atrazine in natural water. Field Anal Chem Technol 2:155–161

Wu C, Siems WF, Asbury GR, Hill HH (1998b) Electrospray ionization high-resolution ion mobility spectrometry / mass spectrometry. Anal Chem 70:4929–4938

Wu C, Siems WF, Klasmeier J, Hill HH (2000) Separation of isomeric peptides using electrospray ionization/high-resolution ion mobility spectrometry. Anal Chem 72:391–395

Wuite GJL, Smith SB, Young M, Keller D, Bustamante C (2000) Single-molecule studies of the effect of template tension on T7 DNA polymerase activity. Nature 404:103–106

Wynne SA, Crowther RA, Leslie AGW (1999) The crystal structure of the human hepatitis B virus capsid. Mol Cell 3:771–780

Xu L, Frederik P, Pirollo KF, Tang WH, Rait A, Xiang LM, Huang W, Cruz I, Yin Y, Chang EH (2002) Self-assembly of a virus-mimicking nanostructure system for efficient tumor-targeted gene delivery. Hum Gene Ther 13:469–481

Yamasaki R, Hoshino M, Wazawa T, Ishii Y, Yanagida T, Kawata Y, Higurashi T, Sakai K, Nagai J, Goto Y (1999) Single molecular observations of the interaction of GroEL with substrate proteins. J Mol Biol 292:965–972

Yarmush ML, Jayaraman A (2002) Advances in proteomic technologies. Annu Rev Biomed Eng 4:349–373

Ying LM, Bruckbauer A, Rothery AM, Korchev YE, Klenerman D (2002) Programmable delivery of DNA through a nanopipet. Anal Chem 74:1380–1385

Yip CM (2001) Atomic force microscopy of macromolecular interactions. Curr Opin Struct Biol 11:567–572

Zal F, Chausson F, Leize E, van Dorsselaer A, Lallier FH, Green BN (2002) Quadrupole time-of-flight mass spectrometry of the native hemocyanin of the deep-sea crab *Bythograea thermydron*. Biomacromolecules 3:229–231

Zhang J, Oettmeier W, Gennis RB, Hellwig P (2002a) FTIR spectroscopic evidence for the involvement of an acidic residue in quinone binding in cytochrome bd from *Escherichia coli*. Biochemistry 41:4612–4617

Zhang ZL, Pang DW, Zhang RY, Yan JW, Mao BW, Qi YP (2002b) Investigation of DNA orientation on gold by EC-STM. Bioconjug Chem 13:104–109

Zheng W, Doniach S (2002) Protein structure prediction constrained by solution X-ray scattering data and structural homology identification. J Mol Biol 316:173–187

Zheng PP, Kros JM, Sillevis-Smitt PA, Theo M Luider Fn FM (2003) Proteomics in primary brain tumors. Front Biosci 8:D451–463

Zhou Y, Karplus M (1999) Interpreting the folding kinetics of helical proteins. Nature 401:400–403

Zhou H, Midha A, Mills G, Thoms S, Murad SK, Weaver JMR (1998) Generic scanned-probe microscope sensors by combined micromachining and electron-beam lithography. J Vacuum Sci Technol B 16:54–58

Zhou H, Mills G, Chong BK, Midha A, Donaldson L, Weaver JMR (1999) Recent progress in the functionalization of atomic force microscope probes using electron-beam nanolithography. J Vacuum Sci Technol A 17:2233–2239

Zocchi G (2001) Force measurements on single molecular contacts through evanescent wave microscopy. Biophys J 81:2946–2953

Index

Abbe's limit of diffraction VII, 135
Aberration
    chromatic 41, 114, 115
    spherical 113, 114
Acyl carrier protein 5, 6
Acyl-coenzyme A binding protein 5, 6
ADC (analog-to-digital converter) 176,
    181, 188, 192
Affinity chromatography 31–33
AFM (atomic force microscope) 121–132,
    142, 143, 145, 147–150, 152, 154, 155,
    158, 159
Alzheimer's disease 131
α-Amylase inhibitor 7
Analog-to-digital converter (ADC) 176,
    181, 188, 192
Anion exchanger 23, 25, 26
Antibodies 32, 152, 161
Antigen 32, 152, 165
Aperture
    grid 176, 178, 179–183
    subwavelength 138–142
apo-Calmodulin 149
Arc repressor 11
Artifacts in atomic force microscopy 125
Atomic force microscope (AFM) 121–135,
ATP (adenosine triphosphate) 135
Attenuated total reflection (ATR) 100, 101
Audibility of microwaves 198–204

Bacillus subtilis 7
Bacterial membrane 88, 147, 149, 154,
    159
Bacterioferritin 5
Bacteriophage
    Φ29 117
    ΦKZ and T4 131
    crystal 158
Bacteriorhodopsin 104, 147, 149

Barnase 9, 11, 12, 48
Barstar 9, 11
BESSY (Berlin Electron Synchrotron
    Storage Ring) 75
Bioagents detection 49–54, 96, 97, 153,
    177, 178, 186, 188, 189, 193–195
Bioelectronic structures 156
Biological agents detection 49–54, 96,
    97, 153, 177, 178, 186, 188, 189,
    193–195
Biomarkers 166
Biophysical nanotechnology VII, 147–163
Botulinum toxin 182
Bovine serum albumin (BSA) 131, 132
Bovine spongiform encephalopathy
    (BSE) 3
Breakaway forces between different
    biomolecules 152
Bright field detection 119
Brookhaven National Laboratory Protein
    Data Bank 2, 5, 7, 8, 14, 18, 19
BSA (bovine serum albumin) 131, 132
BSE (bovine spongiform encepha-
    lopathy) 3

Cantilever 121–135, 143–145, 152, 154
Capacity of ion exchanger 26
Carbon
    fiber AFM tip 127
    nanotube 156, 157
Carboxylic acid 26
Cardiac cells 131
Cation exchanger 23, 25, 26
CBMS (chemical-biological mass
    spectrometer) 49
CCD (charge coupled device)
    area detectors 71–74
    camera 107
CD (circular dichroism) 91

cDNA 167
Cell
    adhesion 130, 131, 147, 150
    lysis 170
Chain topology parameter (*CTP*) 13–16,
    19, 20
Chaperone 20, 33, 131
Chaperonin – *see* chaperone
Charge interaction 5, 19, 23–28, 41–44,
    48, 131, 175
Chemical
    -biological mass spectrometer (CBMS)
        49
    ionization 44
    linkers 2, 152, 155, 168
    proteomics 172
CheY 8
Chromatic aberration 41, 114, 115
Chromatography
    affinity 31–33
    gel filtration 28–31, 171
    gel permeation – *see* gel filtration
    ion exchange 23–28
    of a crude cell extract 47
    plasma 185
    resolution 27, 28
    size exclusion 28–31, 171
Chymotrypsin inhibitor 9–11
Circular dichroism (CD) 91
CJD (Creutzfeldt-Jacob disease) 3
CNS disorders 3, 131, 165
Cold shock protein 7
Collagen fibrils 86, 88
Collector for
    IMS 176, 178–182, 189
    MS 38, 40–42
    protein fractions 26, 28
Collision chamber 45
Compound lens 113, 114
Concentrating protein solutions 34, 35
Contact
    angle measurement 156
    mode 128, 129
    order 13, 15
Continuous-flow method 162
Convolution 66, 67
Convolution theorem 67
Coulomb repulsion 123, 181

Craig counter-current distribution
    apparatus 33, 34
Creutzfeldt-Jacob disease (CJD) 3
Cryocrystallography 84, 85
Crystallization robot 70
Crystallography 59–85
Crystal production 69–71, 158, 159
Crystals suitable for crystallography 69,
    158, 159
CTP (chain topology parameter) 13–16,
    19, 20
Curie point pyrolyzer 49, 51, 193, 195
Cysteine 3, 135, 148
Cytochrome
    b_1 5, 6
    c 135
    P450 135

2D-chromatography 169, 170
2D-electrophoresis 166, 169, 170
Dark field image 119, 120
DEAE (diethyl-amino-ethyl) 23, 25, 26
de Broglie relation 109
Dentin 88
Detection of
    animal activity in jungles 187
    camouflaged organic material 89
    changes in forest structure 97
    chemical activity of proteins 172
    chemicals in solid samples 190
    crystal growth mechanism 158, 159
    drug-protein interactions 154,
        165–167, 170, 171
    environmental pollutants 97, 187
    illicit drugs 187
    intermolecular interactions 152–154,
        168–171
    mass-to-charge ratio 37–57
    membrane pores 144, 154
    non-proteinaceous interaction partners
        165
    organic compounds 175–195
    pesticides 187
    protein-protein interactions 165,
        168–171
    size-to-charge ratio 175
    surface composition 133
    tumor marker 165

ultra-trace chemical and biological
    agents 49–54, 153, 175–195
Detectors for
    electrons 44–46
    ions 37–46, 178, 180–183
    IR radiation 96, 105
    X-rays 72–74, 85, 87, 88
Deuterated triglycine sulfate (DTGS)
    pyroelectric detectors 96
Dichroism
    circular (CD) 91
    linear (LD) 102
Diethyl-amino-ethyl (DEAE) 23, 25, 26
Differentiation between different species
    of bacteria and viruses 52, 53
Diffraction 64–79, 83, 84, 86, 103
Diffraction pattern 64–67, 69, 71–73, 76,
    77, 86
Directional radio 199
Discovery of new gene functions 172
Diseases VII, 1, 3, 17, 50, 88, 131,
    165–167, 209, 211
Dispersion of electron energy 115
Dissecting bacterial surface layers 159
Distribution of conductivity within a
    biomolecule 134
DNA 14, 16, 18, 23, 50, 55, 103, 130,
    131, 135, 150, 151, 153, 154,
    160–162, 166, 167, 169, 173, 174
    microarray 165, 167, 173, 174
    sensor 153, 154
    sequencing 50, 55
Doppler-shift 151
Drift
    channel of an IMS 175–182, 185, 186,
    189
    tube 175, 176, 178–180
    velocity 194
Drug-protein interactions 154, 165–167,
    170, 171
Dynamic force mode 128, 129

Efficiency of folding 17
Elastic dark field image 119
Elastically scattered electrons 115–117,
    119
Elastic modulus 131
Electron

bombardment 44, 51
density 68, 83–85
elastically scattered 115–117, 119
flashes 117
inelastically scattered 108, 115–117,
    119, 120
microscopy (EM) VII, 107–120
multiplier 44–46
nanolithography 141
tomography 117
transport 135
Electrophoresis 55, 166, 169, 170, 185
Electrosmog 197
Electrospray
    deposition 161
    ionization 44, 46, 185
Electrostatic energy analyzer 38, 45, 46,
    115, 116, 120
Electrostatic lens 44, 111, 113, 114, 179,
    190
EM (electron microscope and microscopy)
    VII, 107–120
Energy filter for
    electrons 115, 116, 119, 120
    ions (sector filter) 38, 39, 45, 46
Engineered tags 31, 71, 135
Environmental monitoring 97, 187
Escherichia coli 2, 5
Etched tungsten tip for STM 134
Etching a SNOM probe 138–140
Explosives detection 89, 183, 187

False alarm rate 183, 184
Faraday plate 178, 180–183
Fast performance liquid chromatography
    (FPLC), 26, 28, 46, 48
Feedback loop
    STM 133
    SNOM 136
Fibronectin 7
Field-effect transistor 163
Ω-Filter 116
Flavodoxin 2
Folding
    chaperonin – see chaperone
    mechanism 10, 132
    -related diseases 3
    transition state 9–20

-unfolding forces within single
    protein molecules 132, 147–149
Force
    -extension curve 149
    measurements in single DNA molecules
        150, 151
    measurements in single protein
        molecules 132, 147–149
Forest structure and terrain survey 97
Fourier transform
    infrared (FTIR) spectroscopy 91–96,
        103–105
    mass spectrometer (FTMS) 37, 43, 44
    mathematical method 59–68, 79–83
Fourier transformation
    – see Fourier transform
FPLC (fast performance liquid
    chromatography) 26, 28, 46, 48
FPLC/MS connector 48
Fragmentation of pyrolysates 51, 53
Friction force microscope 143
FTIR (Fourier transform infrared)
    sample cell 96
    spectroscopy 91–96, 103–105
FTMS (Fourier transform mass
    spectrometer) 37, 43, 44
Fumarate reductase 5, 6

Gas chromatography (GC) 46, 47, 184,
    192, 193
GC/IMS (gas chromatography ion mobility
    spectrometry) 184, 193
GC/MS (gas chromatography mass
    spectrometry) 47
Gel
    chromatography – see gel filtration
        chromatography
    filtration chromatography 28–31, 171
    permeation chromatography – see gel
        filtration chromatography
    polyacrylamide 55, 166
Generation of
    amplitude contrast in electron
        microscopy 116
    high voltage 191
Gene therapy 162
Genetically engineered virus 155
Global positioning system (GPS) 96, 97

Glycoprotein 147
Gold nanoparticle 153
GroEL 4, 118
Guard rings 178, 179, 189–191

Hall probe microscope 143
Hanging drop method 70
Heating effects in biological tissue 144
Heavy atom
    replacement 66, 76–83
    salt 76
α-Helix 1, 4, 15, 22, 56, 130, 149, 157
Hemoglobin 5, 6
High frequency octopole ion guide 44
High level expression of proteins 27
High pressure
    IR spectroscopy 103
    liquid chromatography – see HPLC
High resolution mass spectrometer 43, 45
High throughput protein assays 166
High voltage generation 191
Histidine tags 31, 71, 135
holo-Calmodulin 149
HPLC (high pressure liquid
    chromatography) 26, 46
Hybridized DNA 153, 167
Hydrofluoric acid 140
Hydrogen lamp 185, 186
Hydrophilic amino acid sidechains 4, 5,
    19, 156
Hydrophobic amino acid sidechains 1, 4,
    5, 19, 22, 23, 157

Immobilized
    DNA 135, 152, 153, 167
    proteins 135, 152, 168, 170, 172
Impactor 49, 50, 193, 195
IMS (ion mobility spectrometer) 175–195
Inelastically scattered electrons 108,
    115–117, 119, 120
Inelastic dark field image 119, 120
Inert gas for IMS 179
Inertial measurement unit (IMU) 97
Individual intermolecular interactions
    152–154
Infrared
    beam splitter 96
    detectors 96, 105

isotope-edited spectroscopy 102–104
LIDAR (light detection and ranging)
    91, 96, 97
microscope 101, 102
sample cell 96
source 92
spectroscopy 91–105
Ink-jet printing protein spots 168
Insulin 30
Integration host factor (IHF) 130
Interaction proteomics 168–171
Interferogram 93–95
Interferometer 92, 93, 136, 137
Intermolecular
    forces 150–152
    interactions 152–154, 168–171
Inter-residue contacts in protein molecules
    9–17
Inverse Fourier transform 59
Ion
    bombardment 44
    exchange chromatography 23–28
    extractor 174
    fragmentation 45, 51, 53, 54
    guide 44, 48
    mobility spectrometry 175–195
    trap mass spectrometer 37, 39, 44, 53,
        54
Ionization
    chemical 44
    electron bombardment 44, 51
    electrospray 44, 46, 185
    ion bombardment 44
    MALDI 40, 44
IR
    beam splitter 96
    detectors 96, 105
    LIDAR (light detection and ranging)
        91, 96, 97
    microscope 101, 102
    sample cell 96
    spectrometer 92, 93
Isotope-edited FTIR spectroscopy
    102–104

Krypton lamp 185, 186

Lab-on-a-chip technology 165, 173

β-Lactoglobulin 130
Langmuier-Blodgett film 142
Large scale proteomics 165–174
Latex beads 143
LIDAR (light detection and ranging) 91,
    96, 97
Limit of classical optics VII, 135, 136
Linear dichroism (LD) 102
Liquid chromatography 23–34
Loops 1

Macromolecular crowding 17
Magnetic
    lens 112–114, 116
    sector mass spectrometer 37, 38, 46
MALDI (matrix-assisted laser desorption
    ionization) 40, 44
Manipulation of
    DNA 150, 151
    macromolecules 132, 147–157
    protein crystal growth 158, 159
Mars mobile 57
Maser 197, 200–204, 206
Mass spectrometry 37–57, 165, 166, 170,
    171, 174
    Fourier transform 37, 43, 44
    high resolution 43, 45
    ion trap 37, 39, 44, 53, 54
    magnetic sector 37, 38, 46
    quadrupole filter 37, 39
    time-of-flight (TOF) 37, 40–42
Matrix-assisted laser desorption ionization
    (MALDI) 40, 44
MCT (mercury cadmium telluride)
    detectors 96
Mechanism of protein folding 10
Melt-drawing a SNOM probe 139
Membrane
    pore detection 144
    proteins 4, 5, 101, 130, 131, 147, 149,
        154
    sample inlet 177
Mercury cadmium telluride (MCT)
    detectors 96
Michelson interferometer 92, 93, 136, 137
Microarray
    scanner 173
    technology 165–174

Microchannels 173
Microchips 165–174
Microfluid technique 172, 173
Microscope and microscopy
    atomic force (AFM) 121–132, 142,
        143, 145, 147–150, 152, 154, 155,
        158, 159
    bright field 119
    dark field 119, 120
    electron VII, 107–120
    infrared 101, 102
    scanning ion conductance (SICM) 122,
        144
    scanning probe (SPM) 121–145
    scanning thermal (SThM) 144
    scanning transmission electron (STEM)
        118–120
    scanning tunneling (STM) 122, 126,
        133–135
    X-ray 77
Microwave
    auditory effects 198–204
    -based voice transmission 198–204
Microwaves 198–204
Miller indices 67
Millimeterwave camera 198–204
Misfolded protein 3, 4, 33, 71
m^n Mass spectrometry 54
Molecular
    exclusion chromatography – see
        gel filtration chromatography
    redox relays 156
    sieve 179, 180, 184
    weight determination
        chromatographic 31
        mass-spectrometric 37–57
        X-ray spectrometric 89
    wires 155, 156, 157
MOLSCRIPT 2, 4, 5, 7, 8, 18, 118
Molten globule 21, 22
Monochromatic
    electrons 115
    radiation 75, 92, 94, 102, 115
mRNA 167
Myoglobin 5, 6, 30, 105

Nanobiosensor 154
Nanobiotechnology 154, 156, 157, 159

Nanocrystals 155, 158, 160
Nanodissector 159
Nanoelectrodes 160
Nanofabrication 141, 142, 155,
    156, 159–163
Nanolithography 141, 155, 156
Nanometer-sized light source 136–142
Nanopipette 144, 159, 160
Nanotechnology VII, 147–163
Nanotransistors 163
Nanotubes 156, 157
Nanowires 157
Near-field scanning optical microscope
    (NSOM) – see SNOM
Negative
    ion mode 177
    stain 116, 117
Neurodegeneration 165
Neuronal network method 53, 54, 202
Neutron microscope 116
NMR (nuclear magnetic resonance) 4, 174
Nobel prize 109, 134
Noise reduction 59, 183, 190
Non-contact mode 129
Non-proteinaceous interaction partners
    165
NSOM (near-field scanning optical
    microscope) – see SNOM
Nuclear magnetic resonance (NMR) 4, 174

OCT (optical coherence tomography) 98,
    99
Octopole ion guide 44
Open reading frame (ORF) 172
Operating temperature of IMS 178
Optical
    coherence tomography (OCT) 98, 99
    far-field 141
    near-field 141
    tweezers 151
    waveguide 100, 101
Overcoming the classical limits of optics
    VII, 135–142

Particle analysis 49, 50, 193
Peptide sequencing 50, 56, 57
Pesticide detection 187
Phase difference – see phases

Phased array 200, 204, 205, 207, 208
Phases and phase differences
    crystallographic 59–61, 64–66,
        76–83
    determination 66, 76–83
    in optical coherence tomography 99
    in the Fourier transform of a function
        59–61, 64–66
    of an electric field 189
    of electrons in an EM 109, 115–117
    of light in a SNOM shear force
        interferometer 137
    of light in FTIR spectrometers 93, 94
Phosphocarrier protein 8
Photoionization 185, 186
Photon
    detection 73, 74
    flux 186
    shot noise 93
Photosystem 142
Physisorption of protein 168
Piezoelectric scanner 121–125, 133, 137,
    138
Plasma
    chromatography 185
    discharge lamps 186
Plastocyanin 135
Pohlschuh lens 112, 113
Polyacrylamide gel electrophoresis 55, 166
Polychromatic interferogram 94, 95
Polychromator 99
Polymerase 32, 150, 151
Porin 130, 131
Positive ion mode 177
Post-translational changes 50
Preconcentrator 184
Primary
    amine 26
    structure 1
Prion 3, 4
"Prion-only" hypothesis 3
Procarboxypeptidase B 8
Profilometer 122
Propagation of light in a SNOM tip 140
Protein
    crystallography 59–85
    crystal production 69–71
    Data Bank 2, 5, 7, 8, 14, 18, 19

DNA-binding 169
    -drug interactions 154, 165–167, 170,
        171
    engineering 155–158
    folding mechanism 10, 132
    folding simulations 20–22
    folding transition states 9–12
    G 14, 17, 18
    immobilization 135, 152, 168, 170, 172
    L 7
    microarrays 161, 165, 168, 170, 174
    nanoarrays 155, 156, 161
    physisorption 168
    -polymer nanoparticles 162
    -protein interactions 50, 152, 154, 165,
        168–171
    stability 3, 17, 21, 71
    subunits 1, 2, 5
    X-ray crystallography VII, 59–85
Proteomics
    chemical 172
    interaction 167–171
    structural 174
    target discovery 166, 167
PTFE (polytetrafluoroethylene) 96, 177
Pulling a SNOM probe 138, 139
PyMS (pyrolysis mass spectrometry) 49,
    51–54
Pyroelectric detectors 96, 105
Pyrolysis 49, 51–53, 193, 195
Pyrolyzer 49, 51, 193, 195

Quadrupole
    filter mass spectrometer – see
        quadrupole mass spectrometer
    ion trap mass spectrometer
        – see ion trap mass spectrometer
    mass spectrometer 37, 39
Quaternary
    amine 26
    structure 2, 86

RADAR (radio detection and ranging)
    198, 201, 204
Radiation damage by
    electrons 117, 118
    X-rays 65, 71–73, 77
Random coil 1, 22, 105

Reconstitution of
   misfolded protein 33
   molecular sieves 179
Recoverin 149
Refinement 83
Reflectron time-of-flight mass
   spectrometer 40–42
Remote thought transmission 197–209
Repeller plate for
   IMS 177
   MS 40, 42
Resolution of
   AFM 122, 128, 129, 131
   chromatography 24, 47
   classical optical microscope 135
   electron microscope 109, 110, 117–119
   FTIR and IR 91–93, 101, 102, 104
   IMS 180–183, 186, 188, 189, 193
   MS 37, 40, 41, 43, 47
   near-field optical microscope VII, 135,
      136, 138, 142, 143
   SAXS 88
   SICM 144
   STM 134
   Φ-value analysis 9–12
   X-ray crystallography 76
   X-ray microscope 65
Reverse transcriptase 167
RNA 8, 20, 167
Rotating anode generator 71, 74
Rotational evaporation 139
Rupture force for molecular
   complexes 152, 154
   subunits 147

Sample inlet for an IMS 177
Sanger sequencing method 55
SAXS (small angle X-ray scattering)
   85–88
Scanner for microarrays 173
Scanning
   force microscope – see atomic force
      microscope (AFM)
   friction force microscope 143
   Hall probe microscope 143
   infrared microscope 101, 102
   ion conductance microscope
      (SICM) 122, 144

IR spectrometer 92
near-field optical microscope
   (SNOM) 122, 131, 135–143
probe microscopy (SPM) 121–145
thermal microscope (SThM) 144
transmission electron microscope
   (STEM) 118–120
tunneling microscope (STM)
   122, 126, 133–135
SDOCT (spectral domain optical
   coherence tomography 99
Secondary
   amine 26
   structure 1, 4, 5, 7, 9–11, 15, 16, 22,
      56, 91, 105
Sequencing method
   DNA 50, 55
   peptides 50, 56, 57
SH3 domain 7, 9, 11
Shear force 136, 137
β-Sheets 1, 4, 7–9, 15, 16, 22, 105, 126
Shot noise 93
Shutter
   electrode 40, 41
   grid 176–179, 186, 193
   mechanical 92
SICM (scanning ion conductance
   microscope) 122, 144
Silicon nitride
   AFM tips 126, 128
   EM grid 112
   SNOM probe 141
   SThM tip support 144
Silicon wafer IR sample cell 96
Single
   bacteria detection 49–54, 153,
      193–195
   DNA molecule force measurements
      130, 150, 151, 153
   protein molecule force measurements
      147–154
Size exclusion chromatography – see gel
   filtration chromatography
Size-to-charge ratios 175
Skeletal muscle cells 131
Skimmer 46
Small angle X-ray scattering (SAXS)
   85–88

Sniffers 183, 184

SNOM (scanning near-field optical microscope) 122, 131, 135–143

Source-gate channel 163

Spectral domain optical coherence tomography (SDOCT) 99

Spectrometry – *see* spectroscopy

Spectroscopy
electron 115, 116
infrared (IR) and Fourier transform infrared (FTIR) 91–105
ion mobility 175–195
mass 37–57, 165, 166, 170, 171, 174
molecular force 130, 147

Speed of
electrons in an EM 110, 115
ions in an IMS 194
ions in an MS 41

Spherical aberration 113, 114

SPM (scanning probe microscopy) 121–145

STEM (scanning transmission electron microscope) 118–120

SThM (scanning thermal microscope) 144

STM (scanning tunneling microscope) 122, 126, 133–135

Streptavidin-biotin complex 154

Stretch-activated ion channels 159

Structural
characterization by FTIR spectroscopy 102–105
characterization of biomolecules by scanning probe microscopy 131, 132, 134, 147–152, 154, 158, 159
determinants of folding rate constants 12–20
proteomics 174
resolution of proteins by X-ray crystallography 59–85

Structure
characterization by FTIR spectroscopy 102–105
determination by a charged AFM tip 131
determination by X-ray crystallography 59–85
of the native state of proteins 1–8, 118
prediction 20–22

Stylus profilometer 122

Subunits of proteins 1, 2, 5

Subwavelength aperture 138–142

Sulfonic acid 26

Survey of forest structure and terrain 97

Synchrotron 71, 75

T7 polymerase 150, 151

Tags 31, 71, 135

Target discovery in proteomics 166, 167

TEM (transmission electron microscope) 107–118

Tendamistat 7

Tertiary
amine 26
structure 1, 4, 9–11, 91

TGS (triglycine sulfate) detector 105

Therapeutics development 162, 165, 166, 198, 209

Thermocouple probe 144

Thermostable proteins 17–19

Thought
transmission 197–209
transmission apparatus 199–207

Three-dimensional structure
of proteins 1–22, 118

Time-of-flight mass spectrometer 37, 40–42

Titin 1, 147–149

TIR (total internal reflection) 100, 101, 139

TOF (time-of-flight mass spectrometer) 37, 40–42

Total internal reflection (TIR) 100, 101, 139

Transformation of cells 169

Transgenic mice 165

Transition state for protein folding 9–20

Transistor 163

Translation products 165

Transmembrane channel 131, 154

Transmission electron microscope (TEM) 107–118

Trap, optical 151

Triglycine sulfate (TGS) detector 105

Tumor marker 165

Tunneling current 133, 134

Turnip yellow mosaic virus 158

Turns 1
Tweezers, optical 151
Two-dimensional
    chromatography 169, 170
    electrophoresis 166, 169, 170
Two-hybrid system 168, 169
Tyrosine kinase 7

U1A spliceosomal protein 8
Ubiquitin 8
Ultrafiltration 34, 35
Ultrashort electron flashes 117
Ultra-trace detection 49–54, 153,
    175–195
Unfolding of single protein molecules
    147–149

Φ-Value 9, 10
van-der-Waals radii 1, 123, 129, 132
Vector 169
Velocity of
    electrons in an EM 110, 115
    ions in an IMS 194
    ions in an MS 41
Vibrational isolation of an AFM 123, 129,
    130
Viral capsid protein 5, 6
Virtual impactor 49, 50, 193, 195

Virus
    crystallographic analysis 69–71, 77,
        158, 159
    crystals 69–71, 77, 158, 159

Waveguide, optical 100, 101
Wide angle X-ray scattering analysis 85

X-ray
    backscattering 88, 89
    cryogenic system 72
    crystallography 59–85
    detectors 72–74, 85, 87, 88
    diffraction 69–85
    microscope 77
    mirrors 77, 79
    molecular weight determination 89
    poorly scattering crystal 76
    scattering 69–89
    small angle scattering 85–88
    sources 71, 74, 75, 87
    wide angle scattering 85

Yeast
    strains 172
    two-hybrid system 168, 169

Zeolite 179, 184